345
STO

MICROWAVE COMMUNICATIONS

McGRAW-HILL ELECTRICAL AND ELECTRONIC ENGINEERING SERIES

Frederick Emmons Terman, *Consulting Editor*
W. W. Harman and J. G. Truxal, *Associate Consulting Editors*

Ahrendt and Savant · Servomechanism Practice
Angelakos and Everhart · Microwave Communications
Angelo · Electronic Circuits
Aseltine · Transform Method in Linear System Analysis
Atwater · Introduction to Microwave Theory
Bailey and Gault · Alternating-current Machinery
Beranek · Acoustics
Bracewell · The Fourier Transform and Its Application
Brenner and Javid · Analysis of Electric Circuits
Brown · Analysis of Linear Time-invariant Systems
Bruns and Saunders · Analysis of Feedback Control Systems
Cage · Theory and Application of Industrial Electronics
Cauer · Synthesis of Linear Communication Networks
Chen · The Analysis of Linear Systems
Chen · Linear Network Design and Synthesis
Chirlian · Analysis and Design of Electronic Circuits
Chirlian and Zemanian · Electronics
Clement and Johnson · Electrical Engineering Science
Cote and Oakes · Linear Vacuum-tube and Transistor Circuits
Cuccia · Harmonics, Sidebands, and Transients in Communication Engineering
Cunningham · Introduction to Nonlinear Analysis
D'Azzo and Houpis · Feedback Control System Analysis and Synthesis
Eastman · Fundamentals of Vacuum Tubes
Elgerd · Control Systems Theory
Eveleigh · Adaptive Control and Optimization Techniques
Feinstein · Foundations of Information Theory
Fitzgerald, Higginbotham, and Grabel · Basic Electrical Engineering
Fitzgerald and Kingsley · Electric Machinery
Frank · Electrical Measurement Analysis
Friedland, Wing, and Ash · Principles of Linear Networks
Gehmlich and Hammond · Electromechanical Systems
Ghausi · Principles and Design of Linear Active Circuits
Ghose · Microwave Circuit Theory and Analysis
Greiner · Semiconductor Devices and Applications
Hammond · Electrical Engineering
Hancock · An Introduction to the Principles of Communication Theory
Happell and Hesselberth · Engineering Electronics
Harman · Fundamentals of Electronic Motion
Harman · Principles of the Statistical Theory of Communication
Harman and Lytle · Electrical and Mechanical Networks
Harrington · Introduction to Electromagnetic Engineering
Harrington · Time-harmonic Electromagnetic Fields
Hayashi · Nonlinear Oscillations in Physical Systems
Hayt · Engineering Electromagnetics
Hayt and Kemmerly · Engineering Circuit Analysis
Hill · Electronics in Engineering
Javid and Brenner · Analysis, Transmission, and Filtering of Signals
Javid and Brown · Field Analysis and Electromagnetics
Johnson · Transmission Lines and Networks
Koenig and Blackwell · Electromechanical System Theory
Koenig, Tokad, and Kesavan · Analysis of Discrete Physical Systems
Kraus · Antennas
Kraus · Electromagnetics
Kuh and Pederson · Principles of Circuit Synthesis
Kuo · Linear Networks and Systems
Ledley · Digital Computer and Control Engineering

LePage · Analysis of Alternating-current Circuits
LePage · Complex Variables and the Laplace Transform for Engineering
LePage and Seely · General Network Analysis
Levi and Panzer · Electromechanical Power Conversion
Ley, Lutz, and Rehberg · Linear Circuit Analysis
Linvill and Gibbons · Transistors and Active Circuits
Littauer · Pulse Electronics
Lynch and Truxal · Introductory System Analysis
Lynch and Truxal · Principles of Electronic Instrumentation
Lynch and Truxal · Signals and Systems in Electrical Engineering
McCluskey · Introduction to the Theory of Switching Circuits
Manning · Electrical Circuits
Meisel · Principles of Electromechanical-energy Conversion
Millman · Vacuum-tube and Semiconductor Electronics
Millman and Halkias · Electronic Devices and Circuits
Millman and Seely · Electronics
Millman and Taub · Pulse, Digital, and Switching Waveforms
Mishkin and Braun · Adaptive Control Systems
Moore · Traveling-wave Engineering
Nanavati · An Introduction to Semiconductor Electronics
Pettit · Electronic Switching, Timing, and Pulse Circuits
Pettit and McWhorter · Electronic Amplifier Circuits
Pfeiffer · Concepts of Probability Theory
Pfeiffer · Linear Systems Analysis
Reza · An Introduction to Information Theory
Reza and Seely · Modern Network Analysis
Rogers · Introduction to Electric Fields
Ruston and Bordogna · Electric Networks: Functions, Filters, Analysis
Ryder · Engineering Electronics
Schwartz · Information Transmission, Modulation, and Noise
Schwarz and Friedland · Linear Systems
Seely · Electromechanical Energy Conversion
Seely · Electron-tube Circuits
Seely · Electronic Engineering
Seely · Introduction to Electromagnetic Fields
Seely · Radio Electronics
Seifert and Steeg · Control Systems Engineering
Siskind · Direct-current Machinery
Skilling · Electric Transmission Lines
Skilling · Transient Electric Currents
Spangenberg · Fundamentals of Electron Devices
Spangenberg · Vacuum Tubes
Stevenson · Elements of Power System Analysis
Stewart · Fundamentals of Signal Theory
Storer · Passive Network Synthesis
Strauss · Wave Generation and Shaping
Su · Active Network Synthesis
Terman · Electronic and Radio Engineering
Terman and Pettit · Electronic Measurements
Thaler · Elements of Servomechanism Theory
Thaler and Brown · Analysis and Design of Feedback Control Systems
Thaler and Pastel · Analysis and Design of Nonlinear Feedback Control Systems
Thompson · Alternating-current and Transient Circuit Analysis
Tou · Digital and Sampled-data Control Systems
Tou · Modern Control Theory
Truxal · Automatic Feedback Control System Synthesis
Tuttle · Electric Networks: Analysis and Synthesis
Valdes · The Physical Theory of Transistors
Van Bladel · Electromagnetic Fields
Weinberg · Network Analysis and Synthesis
Williams and Young · Electrical Engineering Problems

DIOGENES J. ANGELAKOS
PROFESSOR OF ELECTRICAL ENGINEERING AND COMPUTER SCIENCES
DIRECTOR OF THE ELECTRONICS RESEARCH LABORATORY
UNIVERSITY OF CALIFORNIA, BERKELEY

THOMAS E. EVERHART
PROFESSOR OF ELECTRICAL ENGINEERING AND COMPUTER SCIENCES
UNIVERSITY OF CALIFORNIA, BERKELEY

microwave communications

McGRAW-HILL BOOK COMPANY

NEW YORK   ST. LOUIS   SAN FRANCISCO   TORONTO
LONDON   SYDNEY

MICROWAVE COMMUNICATIONS

Copyright © 1968 by McGraw-Hill, Inc. All Rights Reserved.
Printed in the United States of America. No part of this
publication may be reproduced, stored in a retrieval system,
or transmitted, in any form or by any means, electronic,
mechanical, photocopying, recording, or otherwise, without
the prior written permission of the publisher.
Library of Congress Catalog Card Number 67-26876

01789

1234567890 MAMM 7432106987

1428620

TO HELEN A. AND DORIS E.

preface

An appreciable proportion of present-day communications is transmitted by means of microwave communication systems. A system may be defined as an assemblage of objects united by some form of regular interaction or interdependence to attain an objective. As an example, a microwave communications link is a system whose objective is to provide communication through transmission of voice, codes, television pictures, etc. For an engineer to be able to design or evaluate a system, he must have acquired not only knowledge of existing components and principles but also new concepts and fundamentals on which he can base his understanding of the performance and limitations of new devices as they become available.

The study of microwave communication systems offers an opportunity to present a unified approach to senior students as well as to engineers interested in the communications aspects of electrical engineering. It is the purpose of this book to present a sufficient number of important topics in such detail so that the reader may (1) acquire an understanding in depth of the selected topics, (2) appreciate the engineering approach to the solution of problems and design of components, and (3) become aware of the interrelationship of components in systems.

The book is planned as a text for seniors and is intended for a one-semester or possibly two-quarter course. The length of the course may be adjusted by a choice of one of the typical microwave amplifiers and oscillators of Chapters 3 and 4 and by a selective choice of the material on antennas of Chapter 5. As preparation for a course using this text, the student should be familiar with basic electronic circuitry, such as simple amplifiers, radio-frequency oscillators, simple modulators, and detectors. The elements of Fourier analysis for periodic and aperiodic waveforms should also be understood, as should basic field theory. Modulation (both simple and complex), statistical methods of information processing, as well as special important components and devices not mentioned in the text are very important considerations in microwave communications. These are purposely not included in order that the book retain a textbook flavor and thus avoid the handbook massiveness characterized by the subject matter.

Much of the core material of this book has been used in a course at the Department of Electrical Engineering and Computer Sciences, University of California, Berkeley. Many of us who have taught the course have supplemented, modified, or substituted some of the contents, but the basic concepts have remained as outlined in the text.

Diogenes J. Angelakos
Thomas E. Everhart

contents

Preface ix

1 THE SYSTEM'S CONCEPT *1*

1-1   Introduction 1
1-2   A General Communication System 2
1-3   Specific Systems 4
1-4   The Frequency Spectrum 7
       References 8

2 ELECTROMAGNETIC FIELDS AND POWER *9*

2-1   Introduction 9
2-2   Electromagnetic Wave Propagation 10
       Exercises 2-1 and 2-2 12
2-3   Guided Transverse Electromagnetic Waves 12
       Exercises 2-3 to 2-5 15
2-4   Waveguides 15
       Exercises 2-6 and 2-7 17
2-5   Electromagnetic Energy Propagation 17
       Exercises 2-8 and 2-9 19
       References 19

3 MICROWAVE AMPLIFIERS AND OSCILLATORS 20

3-1   Introduction   20
3-2   High-frequency Limitations of Conventional Tubes   23
      Exercises 3-1 and 3-2   25
3-3   Induced Currents   25
      Exercises 3-3 to 3-7   27
3-4   Transit-time Effects in Space-charge-controlled Tubes   28
      Exercises 3-8 to 3-11   32
          Triode input admittance variation with small transit angles   32
      Exercises 3-12 to 3-14   34
3-5   Klystrons   34
3-6   Velocity Modulation by a Gridded Cavity Gap   38
      Exercises 3-15 and 3-16   41
3-7   The Two-cavity Klystron   41
      Exercises 3-17 to 3-20   48
3-8   The Reflex Klystron   48
          Power and efficiency considerations   54
      Exercises 3-21 and 3-22   55
3-9   Traveling-wave Tubes   56
          Traveling-wave-tube circuits   60   ·   Small-signal traveling-wave-tube analysis   63
      Exercises 3-23 to 3-26   73
      References   73

4 PRINCIPLES OF SOLID-STATE MICROWAVE DEVICES 75

4-1   Introduction   75
4-2   The Tunnel Diode   76
          A Tunnel-diode amplifier   80
      Exercise 4-1   82
4-3   Parametric Amplifiers   82
      Exercises 4-2 and 4-3   84
          The Manley-Rowe relations   85
      Exercise 4-4   87
          Methods of analysis   87
      Exercise 4-5   89
          The parametric up-converter   89
      Exercises 4-6 to 4-9   93
4-4   Masers   93
          The laser   98
      References   101

5 ANTENNAS *103*

- 5-1 *Introduction* *103*
- 5-2 *Types of Antenna Problems* *105*
- 5-3 *Elemental Dipole Antenna* *105*
    Vector and scalar potentials *106* · Far-zone or radiation field *110*
    Exercises 5-1 to 5-3 *112*
- 5-4 *Energy Flow and Power Patterns* *112*
- 5-5 *Wire Antennas* *116*
    Field calculations with the aid of the elemental dipole *117* · Illustrative examples *119* · Circular loop antenna in its lowest mode, uniform current *121*
    Exercise 5-4 *122*
- 5-6 *Antenna Arrays* *124*
    Uniform arrays *127* · Nonuniform arrays *129*
    Exercises 5-5 to 5-9 *131*
    Image techniques *132*
    Exercise 5-10 *134*
- 5-7 *Aperture-type Antennas* *135*
    Exercise 5-11 *142*
- 5-8 *Reflectors for a Given Polar Diagram* *143*
    Exercise 5-12 *144*
- 5-9 *Impedance of Transmitting Antennas* *144*
    Exercise 5-13 *150*
- 5-10 *Receiving Antennas* *150*
    Exercises 5-14 to 5-16 *153*
- 5-11 *Special Reflector-type Antennas Used in Microwave Communication Systems* *154*
    Exercises 5-17 and 5-18 *156*
- 5-12 *Frequency-independent Antennas* *156*
    References *160*

6 PROPAGATION OF RADIO WAVES *162*

- 6-1 *Introduction* *162*
- 6-2 *The Ground-wave Set* *164*
    The space wave *167*
    Exercises 6-1 to 6-4 *171*
- 6-3 *Tropospheric Wave* *172*
    Exercise 6-5 *176*
- 6-4 *Sky-wave Propagation* *176*
    Effective dielectric constant of the ionized region *176* · Reflection from the ionosphere *178*
    Exercises 6-6 to 6-9 *181*

6-5   *Radio-wave Absorption in Atmospheric Gases 182*
        References 183

7 NOISE *186*

7-1   *Introduction 186*
7-2   *Characterization of Noise Sources 189*
7-3   *Thermal Noise 194*
        Exercises 7-1 to 7-4 199
7-4   *Shot Noise 200*
        Exercises 7-5 to 7-9 205
7-5   *Amplitude Characteristics of White Noise 205*
7-6   *Noise Figure and Related Topics 207*
        Exercises 7-10 to 7-12 211
7-7   *Antenna Noise 211*
7-8   *Communication System Noise Model 213*
        Exercises 7-13 and 7-14 214
        References 214

8 SPECIFIC MICROWAVE COMMUNICATION SYSTEMS *215*

8-1   *Introduction 215*
8-2   *Radar System 215*

Choice of frequency 217 · Range of target 218 · The maximum unambiguous range 218 · Radial velocity of target 219 · Radar equation 219 · Radar cross section 220 · Methods of calculation of radar cross section 222 · Maximum radar range 229 · The duty cycle 232 · The pulse width 233 · Pulse recurrence frequency 234 · Scanning rate 234 · Summary 235

Exercises 8-1 to 8-7 235
References 239

Index 241

MICROWAVE COMMUNICATIONS

1

the system's concept

1-1. INTRODUCTION

Communication systems are currently receiving considerable attention from electrical engineers. Recently the first message ever "bounced" off Venus was received on earth. (An electronic computer was necessary to statistically determine that it *had* been received!) Microwave communication links between continents, using artificial satellites as either passive or active repeaters, are being vigorously explored. Military applications of satellite communication systems include satellites carrying TV cameras, which can continuously monitor the portion of the earth (or other planet) below them and periodically transmit the viewed information to a "parent" station. Certainly the field of communication systems is presently an exciting one.

The basic purpose of a communication system is to transmit information from one location to another. A simple illustration is a conversation between two persons. If both speak distinctly (and in a language each understands), information will pass from mouth to ear via sound waves. If the conversation takes place in a quiet room, words spoken by one conversationalist will be heard clearly by both. If 100 pairs of conversationalists are present in the same room and all are talking at once, probably no effective communication will take place. However, if each pair of conversationalists could speak and hear only in a given distinct frequency band and if each pair used a different band, very effective communication would be possible. Frequency multiplexing of this sort is possible by electronic means and demonstrates an important principle in electronic communications—different messages in separate frequency bands can be carried at the same time by a communication channel. The information transmitted per unit time is dependent upon the bandwidth allocated to the signal. The bandwidths allocated to commercial broadcast stations by the Federal Communications Commission (FCC) are listed below.

$B = 10$ kc for AM standard broadcast

$B = 150$ kc for FM standard broadcast

$B = 4.5$ Mc for TV standard broadcast

Generally speaking, the bandwidth is a given fraction of the center frequency, and we have a given gain-bandwidth product. Therefore by using a higher center frequency, the bandwidth in cycles per second is increased, and thus more distinct channels can be amplified with an amplifier at microwave frequencies than at much lower frequencies in the megacycle range.

As will soon become apparent, the overall design of a complex system involves many different components in different frequency bands, and to study the entire system in detail is impractical. Component parts are therefore studied in some detail and then related to the whole system, after a brief look at general and specific typical communication systems.

1-2. A GENERAL COMMUNICATION SYSTEM

One schematic diagram of a general communication system, as given by Shannon,[1]† is shown in Fig. 1-1. The information source might be the

† Raised numbers designate the references at the end of each chapter.

Fig. 1-1 Schematic of a general communication system.

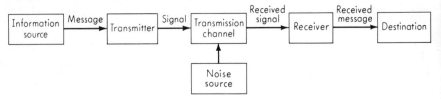

pressure wave arising from speech, a set of punched cards, or a picture to be scanned. The transmitter thus includes the transducer or "coder" which converts the original information into an electrical signal suitable for use in an electronic communication system. It also includes any amplification necessary at this stage and handles the superposition of the original electrical signal representing the intelligence on a high-frequency carrier if the signal is to be radiated or is to be kept separate from other signals on the same channel in a carrier system by frequency-division multiplexing. The coupling element to the transmission channel (e.g., the transmitting antenna) might be considered a part of the transmitter or of the channel itself.

The transmission channel might be a passive transmission line or waveguide or a transmission line with amplifiers (repeaters) in a long-distance telephone transmission but would include the propagation path in a radiated transmission. The receiver includes any coupling element from the transmission channel, such as the receiving antenna; appropriate filters to isolate the desired signal; amplifiers, frequency changers, and demodulators as required to regain a simple electrical signal of desired form; and a transducer or decoder to convert that electrical signal into information of the form finally to be used. Examples of the final decoder would be the loudspeaker of a radio set, the cathode-ray tube in a TV system, or the teleprinter in a Teletype system.

In many cases the original electrical signal is as nearly as possible a direct analog of the original piece of information, as in the case of an audio signal obtained from the sound pressure wave by a microphone used as a transducer. Similarly the original signal may be an analog in which a space coordinate has been converted to a time coordinate in the electrical signal by scanning processes, as in television, facsimile, or radar. On the other hand, it may be a code system in which some pattern of characteristic signals in time (as dots and dashes or characteristic pulse groups) represents given letters, numbers, words, or other entities in the original information source. Such systems of coding may be used to represent any physical function by dividing continuous signals into a set of discrete levels with differences small enough to give the accuracy desired. The receiver "decoder" must convert the final electrical signal back into information of the type desired at the destination (which might of course be different in form from that sent) with as little error as possible arising from the noise. Thus an alternative way of describing the general communication system is as shown in Fig. 1-2. This is a more realistic description

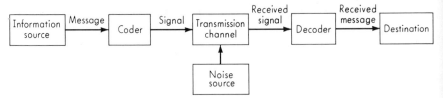

Fig. 1-2 Alternate schematic of a general communication system.

for a number of systems in which some of the transmitter and receiver functions are not clearly separated and in which noise may occur in all parts of the system.

The basic communication system problem is that of using a given transmission channel as effectively as possible. This means that the channel should convey the maximum possible amount of information per unit time consistent with the noise introduced and the error in the received message that can be tolerated. Alternatively the problem may be that of designing the most economical channel to use with the given message consistent with the noise sources and the allowable error rate. The choice of the method of coding as well as the design of all the components in the coder, channel, and decoder is thus of concern to these problems.

1-3. SPECIFIC SYSTEMS

We shall discuss here specific systems (a microwave relay link and a pulsed radar system) in qualitative terms. In the relay link (Fig. 1-3), the transmitting and receiving equipment will be designed for the frequency selected for the given transmission and for the bandwidth required, but it is otherwise standard. Attention is here focused on one of the relay stations as a typical part of the transmission channel, and this is diagramed further in Fig. 1-4. The basic function of the relay link would be satisfied by receiving the wave from the preceding station, amplifying it by the desired amount, and reradiating it toward the next station. However, in such a simple system, isolation between the transmitting antenna and receiving antenna would have to be greater than the gain of the ampli-

Fig. 1-3 Microwave relay system.

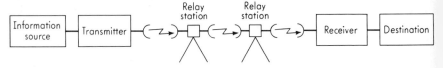

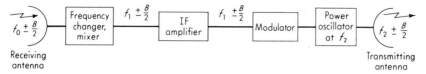
Fig. 1-4 Possible block diagram for one station of microwave relay system.

fier, or the consequent feedback would cause the unit to oscillate or "sing." Thus it is common to change frequency by standard mixing and modulation techniques and send out the information on a slightly different carrier frequency than that received. The bulk of the amplification may be provided at either of the radio frequencies (RF) or at a lower intermediate frequency (IF) (as shown in Fig. 1-4), depending upon the availability of low-noise amplifiers at the various frequencies with bandwidths great enough to handle the desired transmission.

Although the above is a relatively simple system in its basic elements, the system design will have to consider an interplay among a number of factors, of which the following represent examples:

1. In selecting a carrier frequency, the higher this frequency, the greater the bandwidth that can conveniently be obtained for the channel and the smaller the antennas for a given antenna gain but the poorer the efficiency and reliability of available amplifiers. At too high a frequency atmospheric and rain attenuation may also occur along the propagation path.

2. The larger the antennas for a selected carrier frequency, the greater the antenna gain, which could permit less amplification per station or, alternatively, a greater station spacing for a given amplifier. However, large antennas with large wind loadings increase the cost of each station; thus these advantages and disadvantages must be balanced.

3. The design of tall towers or selection of station locations on high mountains or otherwise to take advantage of best propagation possibilities may permit much greater station spacing or cheaper repeater design but may increase the cost of construction or maintenance of the station.

4. The design of all channels with the greatest possible bandwidth would be desirable, as this would permit the multiplexing of many messages or, alternatively, the reception of a given message (with proper coding) under much poorer signal-to-noise ratios and hence with greater station spacings. However, design of very broadband amplifiers and other components may

become expensive, and noise figure may ultimately decrease with bandwidth faster than the potential improvement. Thus a parallel link may be a better solution for increasing channel capacity beyond a certain point.

As a final example, consider the pulsed radar system diagramed in Fig. 1-5. Here the transmitter is turned on for short pulses at regular intervals by the modulator or pulser. These RF pulses go to the antennas through the TR switch, this switch shorting out the path to the receiver while the transmitter is on to prevent damage of the receiver by the large transmitted power. The RF pulse goes out as an electromagnetic wave, and a portion may be reflected back by a suitable target. Thus after a time delay corresponding to the round-trip distance to the target with propagation at substantially the speed of light, the much smaller received pulse is picked up by the antenna and directed by the TR switch to the receiver. This receiver is normally a basic superheterodyne receiver with as low a noise figure as possible. After reception and detection, the pulse is displayed in one of several ways with reference to the pulse-timing circuit and the antenna-directing circuit so that the time delay of the received pulse and the direction of the target can be properly interpreted.

When viewed as a generalized communication system, this is a very interesting one in that the information is added or "coded" on the carrier during the transmission itself. This information is in the *amount of the signal reflected, the time delay of the received reflected pulse with respect to that sent out, the doppler frequency shift of the reflected signal if the target is moving, and possibly the change in polarization of the signal upon reflection.* In "decoding," the problem is that of obtaining desired quantities such as range, range rate, direction, and relative sizes from this "coding" in the presence of the interfering noise introduced along the transmission path

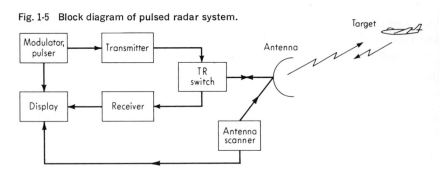

Fig. 1-5 Block diagram of pulsed radar system.

and in the first stage of the receiver. However, one advantage makes the problem easier. The form of the received pulse is known, and correlation-type comparisons with the original transmitted pulse make very sensitive detection methods possible with signals only a small fraction the size of the noise.

In the design problem, interrelations among transmitted frequency, antenna size, transmitted power, receiver noise figure, and characteristics of the propagation path are again important, much as in the example of the microwave relay link. Pulse width and repetition rate offer two additional variables of the problem.

1-4. THE FREQUENCY SPECTRUM

Since so many of the electronic communication systems with which we are concerned make use of the space propagation path and since a certain bandwidth is required for each transmission, it is clear that the frequency spectrum is an international resource, and it is so treated. The international body concerned is the International Telecommunications Union (CCIR). In this country allocations are handled by the FCC, which receives technical advice from the radio industry through the Joint Technical Advisory Committee (JTAC). Although the allocations are set by considering in a general way the bandwidth requirements and best propagation characteristics for a given service, compromises between the requirements of all services must be made. Thus it is clear that in choosing the frequency for a given design, the best possible frequency for that transmission may not be available.

The general designations of frequency ranges are given below:

FREQUENCY f	WAVELENGTH	DESIGNATION
Below 30 kc	Above 10 km	VLF (very low frequency)
30–300 kc	10–1 km	LF (low frequency)
0.3–3 Mc	1–0.1 km	MF (medium frequency)
3–30 Mc	100–10 m	HF (high frequency)
30–300 Mc	10–1 m	VHF (very high frequency)
300–3,000 Mc	1–0.1 m	UHF (ultrahigh frequency)
3–30 Gc	100–10 mm	SHF (superhigh frequency)
30–300 Gc	10–1 mm	EHF (extremely high frequency)†

† If one has to go higher, there is always the British-proposed EHFI for "extremely high frequency indeed!"

Of these designations, the terms UHF and VHF are most often used in their exact contexts, but others are often used more loosely. The term *microwave frequencies* is very commonly used for those wavelengths measured in centimeters, roughly from 1,000 Mc upward in frequency (30-cm wavelength and shorter) and usually including the range of millimeter wavelengths. It is clear that the microwave ranges provide potentially tremendous bandwidths compared with the lower frequency ranges, but the propagation characteristics (to be discussed in Chap. 4) are special and have so far limited the usefulness to special services, although the advent of "scatter" and "meteor burst" propagation techniques are now extending those services. Lack of tubes and components eliminates some possibilities at the highest frequency ranges. Although complete selections of components are available at a number of places in the millimeter-wave range (chiefly in the vicinity of 8.6 and 4.3 mm), efficiency or power output of oscillators and noise figures of amplifiers or mixers are inferior to those in the centimeter-wave region at this date.

Military designations within the microwave range are:

DESIGNATION	FREQUENCY RANGE, Mc
P band	225–390
L band	390–1,550
S band	1,550–5,200
X band	5,200–10,900
K band	10,900–36,000
Q band	36,000–46,000
V band	46,000–56,000

At this point the study of some of the important frequency allocations[2] might be in order with some speculation concerning the reasons for them. A JTAC publication[3] also gives some recommendations for more ideal assignments if one wishes to follow up the matter of spectrum utilization.

REFERENCES

1 SHANNON, C., and W. WEAVER: "The Mathematical Theory of Communication," p. 5, University of Illinois Press, Urbana, Ill., 1949.

2 "Reference Data Radio Engineers," 4th ed., pp. 10–15, International Telephone and Telegraph Corporation, New York, 1956.

3 Joint Technical Advisory Committee (IRE-RTMA): "Radio Spectrum Conservation," pp. 149–150, McGraw-Hill Book Company, New York, 1952.

2

electromagnetic fields and power

2-1. INTRODUCTION

Just as wires are used to connect components in a low-frequency electrical circuit, so transmission lines and waveguides are used to connect component parts of a microwave communication system. When the dimensions of the circuits of a system become comparable to the wavelength of the energy being utilized in that system, energy can be radiated. Except for the case of antennas, such radiation is undesirable for two reasons: first, it represents a loss of useful power, and second, it can lead to undesirable feedback effects which alter the performance of the system and sometimes result in oscillations. It is for these reasons that microwave energy is always conveyed from one part of a system to another via a transmission line or through a waveguide, either of which shields the region carrying energy from the external environment. In certain cases microwave energy is conveyed from one part of a system to another through free space; i.e., the energy is purposefully radiated from an antenna. It propagates through space as an electromagnetic wave; in a radar system, for example, it is reflected back to the transmitting antenna, and in a microwave relay link it is received by the receiving antenna. In either case electromagnetic wave propagation must be treated as part of the system. Basic to the understanding either of transmission lines and waveguides or of free-space electromagnetic-wave propagation is an understanding of electromagnetic theory based on Maxwell's equations. It is assumed that the reader is familiar with the subject of electromagnetic theory to the level covered in standard references.[1] In this chapter we shall briefly review aspects of electromagnetic theory which will be useful to the concepts presented in succeeding chapters. In addition, certain basic definitions which are essential to the understanding of amplifiers as used in microwave systems will be presented.

2-2. ELECTROMAGNETIC WAVE PROPAGATION

Electromagnetic wave propagation is based on the solution of Maxwell's equations which relate the electric flux density **D**, magnetic flux density **B**, the electric field **E**, and the magnetic field **H** to the conduction or convection current density **i** and the charge density ρ. Maxwell's equations are stated as follows:

$$\nabla \times \mathbf{E} = -\frac{\partial \mathbf{B}}{\partial t} \tag{1}$$

$$\nabla \times \mathbf{H} = \mathbf{i} + \frac{\partial \mathbf{D}}{\partial t} \tag{2}$$

$$\nabla \cdot \mathbf{B} = 0 \tag{3}$$

$$\nabla \cdot \mathbf{D} = \rho \tag{4}$$

In a linear, isotropic, and homogeneous material, **D** is proportional to **E** and **B** is proportional to **H**. The proportionality constants are the permittivity (or dielectric constant) ϵ and the permeability μ. Thus

$$\mathbf{D} = \epsilon \mathbf{E} \tag{5}$$

$$\mathbf{B} = \mu \mathbf{H} \tag{6}$$

In nonlinear media, ϵ and μ depend on **E** and **H**; in anisotropic media, ϵ and μ differ depending on the direction of the field, and a tensor must be used to represent either or both. In nonhomogeneous media, ϵ and μ vary from point to point or region to region.

The wave equation for linear, isotropic, homogeneous, charge-free, and current-free media results directly from Eqs. (1) to (6), with **i** and ρ set equal to zero. By taking the curl of (1) and substituting in (2), using (6), a relationship involving only the electric field is obtained. Thus

$$\nabla \times \nabla \times \mathbf{E} = -\mu \frac{\partial}{\partial t}(\nabla \times \mathbf{H}) = -\mu\epsilon \frac{\partial^2 \mathbf{E}}{\partial t^2} \tag{7}$$

Using the auxiliary vector relationship given as

$$\nabla(\nabla \cdot \mathbf{E}) - \nabla^2 \mathbf{E} = \nabla \times \nabla \times \mathbf{E} \tag{8}$$

the vector wave equation for the electric field is obtained. Using an analogous process for the magnetic field, the vector wave equations for the magnetic field may also be obtained. These are given as

$$\nabla^2 \mathbf{E} = \mu\epsilon \frac{\partial^2 \mathbf{E}}{\partial t^2}$$
$$\nabla^2 \mathbf{H} = \mu\epsilon \frac{\partial^2 \mathbf{H}}{\partial t^2} \tag{9}$$

Solution of the electromagnetic wave equations subject to the appropriate boundary conditions determines the propagation of the electromagnetic energy to any particular system of conductors. The simplest solution of the wave equation yields a plane wave of infinite extent. If the electric vector is in the $+x$ direction and propagation is in the $+z$ direction, the electric field may be written

$$\mathbf{E} = \mathbf{a}_x E_x e^{j(\omega t - kz)} \tag{10}$$

By substitution of (10) into Maxwell's equations, the relationship between the propagation constant k and the radian frequency ω may be obtained. Thus

$$k^2 = \omega^2 \mu\epsilon = \frac{\omega^2}{c^2} \tag{11}$$

where c is the velocity of light in the media. It also follows from substitution of (10) into (1) that the vector magnetic field is given by

$$\mathbf{H} = \mathbf{a}_y \frac{E_x}{\eta} e^{j(\omega t - kz)} \tag{12}$$

where $\eta = (\mu/\epsilon)^{\frac{1}{2}}$ is the intrinsic impedance of the media. In free space $\eta_0 = (\mu_0/\epsilon_0)^{\frac{1}{2}} = 377$ ohms, and $c = (\mu_0\epsilon_0)^{-\frac{1}{2}} \approx 3 \times 10^8$ m/sec.

From the above it is seen that a plane wave with an electric vector in the $+x$ direction has a magnetic vector in the $+y$ direction which is proportional to the electric vector in amplitude at each instant of time

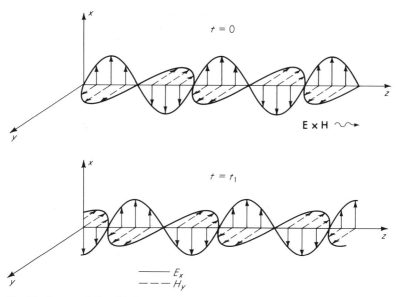

Fig. 2-1 Space relationship between E and H in a plane TEM wave.

and point in space. A diagram showing some of the features of such a plane wave is shown in Fig. 2-1. The pattern shown in Fig. 2-1 moves in the $+z$ direction at the velocity of light. Note that in this plane wave both the electric and the magnetic fields are perpendicular to the direction of propagation of energy and to each other.

EXERCISES

2-1 The electric field **E** of a plane wave traveling in a direction **n** may be written $\mathbf{E} = \mathbf{E}_0 \exp j(\omega t - \mathbf{k} \cdot \mathbf{r})$, where $\mathbf{E}_0$ is a constant and $\mathbf{k} = \omega \sqrt{\mu\epsilon}\, \mathbf{n}$, where **n** is a unit vector. Verify that this is a plane wave, and determine the restrictions on $\mathbf{E}_0$. Find **H** in terms of the quantities already given.

2-2 In the above exercise, $\mathbf{E}_0$ is both a vector and a phasor. If $\mathbf{E}_0 = |E_0|(\mathbf{a}_x + j\mathbf{a}_y)$, determine the locus of **E** as a function of time. What is the projection of **E** on the z plane?

2-3. GUIDED TRANSVERSE ELECTROMAGNETIC WAVES

Ideally electromagnetic energy is guided by perfect conductors which have no losses. In practice no conductor presently used extensively is a perfect conductor, and therefore a certain amount of energy is lost in the transmission line. However, in this section we shall assume the conductors are perfect. If **n** is the unit normal pointing outward from the surface of a perfect conductor, the relationships between **n** and the electromagnetic field quantities at the boundary of the conductor are given by the following relationships:[2]

n · D = ρ_s	$D_n = \rho_s$	(13)
n × E = 0	$E_t = 0$	(14)
n × H = **J**	$\|H_t\| = \|J\|$	(15)
n · B = 0	$B_n = 0$	(16)

In Eq. (13) ρ_s is the surface-charge density at the surface of the perfect conductor on which the lines of **D** originate, and in Eq. (15) **J** is the surface-current density which induces (or is induced by) the magnetic field **H**, which is parallel to the boundary of the conductor. Equation (14) results from the continuity of the tangential component of the electric field vector **E** across a boundary and the fact that within a perfect conductor there can be no electric field (if there were, there would of necessity be an infinite current density). Equation (16) follows from Eq. (14) by using (1). The surface-current density **J**, which is expressed in units of amperes per meter in (15), may be considered as the limiting case of a current density **i** which extends a small distance d into the conductor as $d \to 0$ and $\mathbf{i} \to \infty$. These boundary conditions for a perfect conductor are also useful for metals with high but finite conductivity, such as copper or silver, for time-varying fields (especially when the rate of variation is at a microwave frequency).

Consider next a plane electromagnetic wave which propagates between two very wide, infinitely long metallic sheets whose planes are perpendicular to the x axis and separated by a distance x_0, as shown in Fig. 2-2a. Surface currents and charges will accompany the wave as it propagates in the z direction between these two plates. It is left as an exercise to show that the wave described by (10) and (12) satisfies the boundary conditions previously stated, if the fields exist only between the metal sheets, i.e., for $0 \leq x \leq x_0$. Since the wave is intimately connected with the surface currents and charges which flow in the sheets, these sheets can be perturbed somewhat from the position shown in Fig. 2-2a and the wave will still propagate if the perturbation is gradual (Fig. 2-2b). For example, the thin sheets could be flexed such that they curved around a line parallel to the z axis; by bending them into a full circle, two coaxial

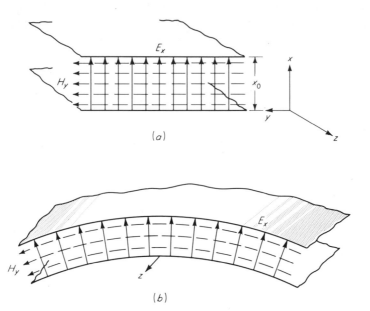

Fig. 2-2 (a) Parallel plane propagation; (b) curved "parallel" plane propagation.

tubes with a difference in radius small compared with the radius of either tube could be obtained. Then if the inner tube were shrunk to a much smaller size and (since there is no field inside this tube) the air inside were replaced with solid metal, a coaxial line would result. Although this reasoning is very qualitative, it suggests that a transverse electromagnetic wave might be expected to propagate on a coaxial line. The fields of this wave can be expressed as shown below.

$$E_r = E_a \frac{a}{r} e^{j(\omega t - kz)}$$
$$\eta H_\phi = E_a \frac{a}{r} e^{j(\omega t - kz)}$$
(17)

From (17) the electric field is in a radial direction and the magnetic field in an azimuthal direction. Both are mutually perpendicular and proportional to each other, the proportionality constant being the intrinsic impedance of the medium. Both electric and magnetic fields are in phase with each other, as was demonstrated for the plane wave discussed above. It is left as an exercise to demonstrate that the fields expressed in (17) do satisfy Maxwell's equations in cylindrical coordinates and as an additional exercise to evaluate the field magnitude E_a in terms of the voltage which exists across the line.

EXERCISES

2-3 Demonstrate that the fields in (17) do satisfy Maxwell's equations in cylindrical coordinates.

2-4 Define voltage as the integral of the electric field at a constant position in z, and at a given instant of time, determine the relationship between E_a given in (17) and the voltage between the inner and outer conductors of the coaxial line.

2-5 Determine the current which flows in the inner conductor of the coaxial line in terms of the voltage found above, and thus obtain the characteristic impedance of the line, which is the ratio of the voltage to the current.

2-4. WAVEGUIDES

Microwave energy is often guided from one part of a system to another through hollow pipes called waveguides. Commonly used waveguides are either rectangular or circular in cross section. In a coaxial line a dielectric is needed to support the inner conductor. This dielectric has associated with it losses which will attenuate electromagnetic waves propagating along the line. Such is not the case in hollow waveguides, which are generally air-filled but in special applications may be either evacuated or pressurized to increase power-handling capability. Unlike the coaxial line, however, which will propagate energy to zero frequency, waveguides have a finite cutoff frequency below which energy will not propagate but is reactively attenuated. Like the coaxial line, waveguides have higher-order modes which can also propagate in them if the frequency is sufficiently high. Thus their useful frequency band in which only a single mode will propagate is considerably less than that of a coaxial line.

The solution to Maxwell's equations applicable in waveguides shows that two classes of waves exist, those which are transverse electric, i.e., have no electric field in the axial direction, and those which are transverse magnetic, i.e., have no magnetic field in the axial direction. If the axial magnetic field is correctly specified for the waveguide in the transverse electric case, then all other fields can be derived by proper application of Maxwell's equations. Using the coordinate system specified in Fig. 2-3, the electric and magnetic fields for the TE_{mn} modes may be written as

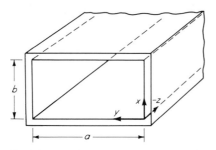

Fig. 2-3 Coordinate system for rectangular waveguide.

follows:

$$H_z = A \cos \frac{m\pi y}{a} \cos \frac{n\pi x}{b}$$
$$E_x = j \frac{\eta m\pi f}{k_c a f_c} A \sin \frac{m\pi y}{a} \cos \frac{n\pi x}{b}$$
$$E_y = -j \frac{\eta n\pi f}{k_c b f_c} A \cos \frac{m\pi y}{a} \sin \frac{n\pi x}{b} \quad (18)$$
$$H_y = +\frac{E_x}{Z_{TE}} \quad H_x = -\frac{E_y}{Z_{TE}}$$

In the above equations the propagation factor $\exp j(\omega t - \beta z)$ is understood and the following definitions have been used:

$$k_c = \frac{2\pi}{\lambda_c} = 2\pi f_c \sqrt{\mu\epsilon} = \left[\left(\frac{m\pi}{a}\right)^2 + \left(\frac{n\pi}{b}\right)^2\right]^{\frac{1}{2}} \quad (19)$$

$$\beta = \omega\sqrt{\mu\epsilon}\left(1 - \frac{f_c^2}{f^2}\right)^{\frac{1}{2}} \quad (20)$$

$$Z_{TE} = \eta\left(1 - \frac{f_c^2}{f^2}\right)^{-\frac{1}{2}} \quad (21)$$

The lowest-order mode which can propagate in a rectangular waveguide is the TE_{10} mode, whose fields are listed below:

$$H_z = j\frac{E_0}{\eta}\frac{\lambda}{2a}\cos\frac{\pi y}{a}$$
$$E_x = E_0 \sin\frac{\pi y}{a}$$
$$H_y = +\frac{E_0}{Z_{TE}}\sin\frac{\pi y}{a} \quad (22)$$
$$H_x = E_y = E_z = 0$$

In this mode the electric field varies in only one transverse direction. Cutoff in this guide occurs when the transverse dimension a is one-half of a free-space wavelength; that is,

$$\lambda_c = 2a = \frac{1}{f_c \sqrt{\mu\epsilon}} \tag{23}$$

Physically, cutoff occurs at a transverse resonance in the guide. By referring to Eqs. (22), note that $H_y = 0$ at cutoff ($Z_{TE} \to \infty$) and that H_z and E_x are in time quadrature. At cutoff the energy is stored first in the electric field and then in the magnetic field, both of which are represented by transverse standing waves in the waveguide. This physical interpretation of cutoff is often very helpful when it is necessary to calculate the cutoff wavelength of guides partially filled with dielectric, etc.

EXERCISES

2-6 The axial electric field for a transverse magnetic wave may be written

$$E_z = A \sin\left(\frac{m\pi y}{a}\right) \sin\left(\frac{n\pi x}{b}\right)$$

with $\exp j(\omega t - \beta z)$ understood. Using Maxwell's equations, derive the fields for the transverse magnetic waves in a rectangular waveguide from this axial electric field, and write the equations which correspond to (18) to (21).

2-7 A rectangular cavity is formed by shorting a rectangular waveguide at the plane $z = 0$ and the plane $z = d$; if the guide dimensions are such that $d > a > b$, determine the lowest resonant frequency of this cavity.

2-5. ELECTROMAGNETIC ENERGY PROPAGATION

In systems calculations we are most often interested in the energy flow from one part of the system to another. In microwave systems this energy is in the form of electromagnetic waves; therefore we must be able to calculate the energy flow in an electromagnetic wave. Let us consider a general case in which the dielectric constant and permeability may be functions of space but not of time. Electromagnetic field quantities satisfy Maxwell's equations and the auxiliary relations as given in (1) to (6). We begin by writing the vector identity

$$\nabla \cdot (\mathbf{E} \times \mathbf{H}) = \mathbf{H} \cdot \nabla \times \mathbf{E} - \mathbf{E} \cdot \nabla \times \mathbf{H} \tag{24}$$

By substituting (1) and (2) in the right-hand side,

$$\nabla \cdot (\mathbf{E} \times \mathbf{H}) = -\mathbf{H} \cdot \frac{\partial \mathbf{B}}{\partial t} - \mathbf{E} \cdot \frac{\partial \mathbf{D}}{\partial t} - \mathbf{E} \cdot \mathbf{i} \qquad (25)$$

Note that (25) is of the appropriate form since power is related to the product of voltage and current (or the analogous field quantities of electric and magnetic field). Realizing that ϵ and μ are constant in time, we may write (25) as

$$\nabla \cdot (\mathbf{E} \times \mathbf{H}) = -\frac{\partial}{\partial t}\left(\frac{1}{2}\mu H^2\right) - \frac{\partial}{\partial t}\left(\frac{1}{2}\epsilon E^2\right) - \mathbf{E} \cdot \mathbf{i} \qquad (26)$$

where the quantity $(\frac{1}{2}\mu H^2)$ is identified as the magnetic energy density per unit volume, the quantity $(\frac{1}{2}\epsilon E^2)$ is identified as the electric energy density per unit volume, and the quantity $\mathbf{E} \cdot \mathbf{i}$ is a power density term. In the case of an ohmic material where $\mathbf{i} = \sigma \mathbf{E}$, the power density term represents the power loss in the material. In the case of free charges in vacuum, this term represents the power which goes into the kinetic energy of the charges being accelerated by the field. Next we integrate (26) over the volume, making use of Gauss' integral theorem to obtain the macroscopic relationship

$$\int_V \nabla \cdot (\mathbf{E} \times \mathbf{H}) \, dv = \oint_S (\mathbf{E} \times \mathbf{H}) \cdot d\mathbf{S} = -\int_V \left[\frac{\partial}{\partial t}\left(\frac{1}{2}\mu H^2\right) + \frac{\partial}{\partial t}\left(\frac{1}{2}\epsilon E^2\right) + \mathbf{E} \cdot \mathbf{i}\right] dv \qquad (27)$$

The left-hand side of this equation represents the integral of $\mathbf{E} \times \mathbf{H}$ over a *closed* surface, where the elemental vector area $d\mathbf{S}$ is perpendicular to and directed outward from the closed surface. The right-hand integral of (27) is seen to be the time rate of change of magnetic energy, the time rate of change of electric energy, and the power dissipated inside the volume. Since the right-hand side of the equation is negative, it follows that the left-hand side must be associated with the energy flow outward across the closed surface. Hence the term $\mathbf{E} \times \mathbf{H}$ is identified as the power flow per unit area in an electromagnetic wave and is identified as the Poynting vector,

$$\mathbf{P} \equiv \mathbf{E} \times \mathbf{H} \qquad (28)$$

Note that whereas it is often useful to identify **P** with the power flow or energy flow per unit area, nevertheless the derivation shows it has true physical significance only when it is integrated over a closed surface.

Care must be exercised when computing the Poynting vector from phasor fields. The instantaneous value of **P** for the plane wave of Sec. 2-2 is obtained by taking the vector product of the *real part* of the electric and magnetic fields. Thus

$$\mathbf{P} = |E_x| \cos(\omega t - kz - \phi) \frac{|E_x|}{\eta} [\cos(\omega t - kz - \phi)]\mathbf{a}_z \qquad (29)$$

where $E_x = |E_x|e^{-j\phi}$. The time-average value of **P** is

$$\bar{\mathbf{P}} = \mathbf{a}_z \frac{|E_x|^2}{2\eta} \qquad (30)$$

EXERCISES

2-8 Calculate the power in a TE_{10} rectangular waveguide mode with fields given by (22).

2-9 If the vector electric and magnetic time-varying fields are represented by complex magnitudes (phasors), show that the time-average power density in the field is correctly represented by $\frac{1}{2}$ Re (**E** × **H***), where **H*** denotes the complex conjugate of **H**.

REFERENCES

1 RAMO, S., J. R. WHINNERY, and T. VAN DUZER: "Fields and Waves in Communication Electronics," chaps. 1–6, John Wiley & Sons, Inc., New York, 1965.
2 Ref. 1, p. 193.

3

microwave amplifiers and oscillators

3-1. INTRODUCTION

In this chapter we shall be interested in the generation and amplification of power at microwave and higher frequencies, i.e., at frequencies above 1 Gc. The early attempts to generate coherent electrical power at microwave frequencies were extensions of space-charge–limited vacuum-tube technology, using triodes, tetrodes, and pentodes. Conventional tubes of this sort suffered from two basic limitations: parasitic energy-storage circuit elements, such as interelectrode capacitance and lead inductance, and finite transit time of electrons between electrodes. By incorporating the grid-cathode and grid-plate capacitance into resonant cavities and by reducing electrode spacings, especially the grid-cathode spacing, these effects were minimized and tubes such as the "lighthouse" tube shown in Fig. 3-1 resulted. The Western Electric 416 series, with grid-cathode spacing of less than 0.001 in. and operating limit of about 4 Gc, represents the limits of technology in pushing this type of tube to higher frequencies.

More generally useful in this frequency range are those tubes which utilize transit-time effects for their operation. The first of these was the two-cavity klystron. It was followed by the reflex klystron, the magne-

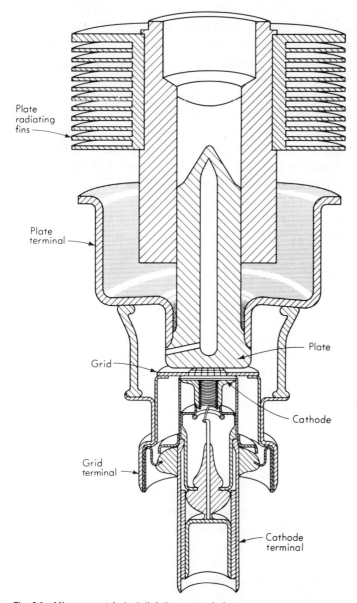

Fig. 3-1 Microwave triode ("lighthouse" tube).

tron, and the family of traveling-wave tubes (TWTs), including the forward-wave amplifier (FWA), the backward-wave amplifier and oscillator (BWA and BWO), and the linear magnetron amplifier. Although it is beyond our scope to discuss all these devices in detail, we shall point out common principles used in many of them and analyze one or two in some detail to illustrate the methods which one must use.

Another class of devices becoming increasingly important at microwave and higher frequencies is the solid-state devices. These can be broken into three subgroups: those which operate by the stimulated emission of radiation (quantum-electronic devices), those which utilize junction effects (semiconductor devices), and those which utilize nonreciprocal effects (ferrite and ferroelectric devices). The first group includes the *maser* (*m*icrowave *a*mplification by the *s*timulated *e*mission of *r*adiation) and the *laser* (*l*ight *a*mplification by the *s*timulated *e*mission of *r*adiation). It seems probable that amplifiers and oscillators based on the maser principle will be useful over the frequency spectrum from microwave frequencies to ultraviolet light frequencies and higher. The second group includes transistors, varactor diodes used in parametric amplifiers and frequency multipliers, and tunnel diodes, which have a fundamental negative resistance. These devices are useful at microwave frequencies and beyond because the dimension of their junction is short compared with a wavelength (being of the order of 0.001 in. in a varactor diode down to about 50 Å for a tunnel diode) and the transit-time problem either is eliminated (varactor) or is unimportant (tunneling is virtually an instantaneous phenomenon). The third group finds its primary application in nonreciprocal elements, such as isolators, circulators, and gyrators. Although these are not amplifying devices, they make possible certain types of amplifiers when used in conjunction with other devices, such as the varactor diode. A fourth group of solid-state amplifiers which may become important as solid-state technology develops would utilize a bulk interaction between electric waves and electrons in the solid. One example of this type of device is the acoustical-wave amplifier.

At the highest radio frequencies presently used (wavelengths of a few millimeters), the tubes listed above are again approaching fundamental limits, since their dimensions are related to wavelength and their size becomes so small that they are difficult to fabricate. In this case they will dissipate little power, and necessary dimensional tolerances for both the structure and the electron beam become difficult to attain. As a result, noise figure and efficiency both worsen. The solid-state junction devices are bothered by lead inductance and losses, such that they become more difficult to use and their efficiency decreases. On the other hand, solid-state resonance devices such as the maser and laser become more efficient and useful at higher frequencies.

3-2. HIGH-FREQUENCY LIMITATIONS OF CONVENTIONAL TUBES

Conventional triodes, tetrodes, and pentodes are less useful active circuit elements at frequencies above 100 Mc to 1 Gc (depending upon the tube) because parasitic capacitance between tube electrodes and leads and parasitic lead inductance become important at these frequencies. As the frequency increases, the time required for an electron to move from cathode to grid becomes a larger and larger fraction of the period of the exciting voltage; these transit-time effects cause the input resistance of the tube to decrease (in the common-cathode connection).

To better understand parasitic inductance and capacitance effects, consider the following example. Figure 3-2a shows a conventional pentode with a rather long cathode lead which has a parasitic inductance L_k.

Fig. 3-2 (a) Pentode circuit with parasitic cathode inductance; (b) equivalent circuit of (a); (c) more complete equivalent circuit of (a), showing inherent tube capacitances and possible parasitic lead inductances.

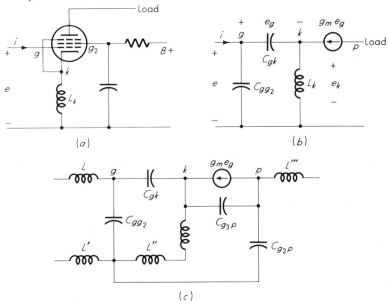

Figure 3-2b shows the equivalent circuit of this pentode, assuming internal capacitance and the cathode inductance are the only parasitic elements. The primary effect of the cathode inductance is to alter the input admittance to the tube. This can be calculated from the equivalent circuit (assuming the grid-plate capacitance is negligible) as follows: The input voltage e may be written

$$e = e_g + e_k \tag{1}$$

and since $\omega L_k \ll 1/\omega C_{gk}$, the cathode voltage is very nearly

$$e_k = j\omega L_k g_m e_g \tag{2}$$

The current i is given by

$$i = j\omega C_{gg_2} e + j\omega C_{gk} e_g \tag{3}$$

and from (1) and (2) we determine that the grid voltage e_g is

$$e_g = e(1 + j\omega L_k g_m)^{-1} \tag{4}$$

Unless the cathode lead is very long or it has a very small diameter or the frequency is very high, the inequality $\omega L_k g_m \ll 1$ is almost always satisfied, which allows us to write

$$i = (j\omega C_{gg_2} + j\omega C_{gk} + \omega^2 L_k C_{gk} g_m)e = Ye \tag{5}$$

giving the input admittance as

$$Y = j\omega(C_{gk} + C_{gg_2}) + \omega^2 L_k C_{gk} g_m \tag{6}$$

The imaginary part of the admittance is the usual input capacitance; the real part is a conductance which increases with the square of the frequency. Although it is small compared with the reactive component, under our assumptions, it nevertheless can become large enough to seriously load the input circuit at high enough frequencies. Note also that the effect of the capacitive susceptance also increases directly with frequency; thus a small variation in wiring capacitances can often appreciably alter resonant frequencies in the 50- to 500-Mc range.

Lest the reader think that we have now exhaustively considered high-frequency effects in pentodes, he is urged to study and verify the equivalent circuit of Fig. 3-2c, which includes several other parasitic elements neglected in the above analysis.

EXERCISES

3-1 Determine the input impedance of a triode with a long cathode lead to ground and a long plate lead to the plate load resistor R_L, where the output voltage e_o is measured. Express your answer in terms of parasitic elements and the tube gain.

3-2 Compare the change in input admittance due to the parasitic inductance for the tube and transistor circuits shown in Fig. 3-3. Which circuit is more sensitive to parasitics of this sort?

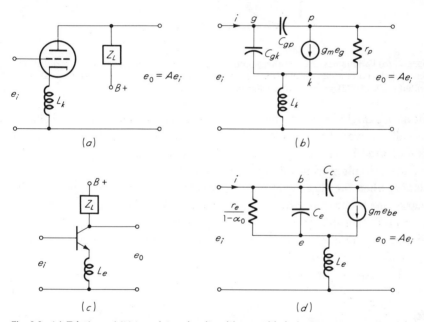

Fig. 3-3 (a) Triode and (b) transistor circuits with parasitic inductances.

3-3. INDUCED CURRENTS

The mechanism of current induction by moving charges was developed by Shockley[1] and Ramo.[2] As this point is so basic to the operation of high-frequency tubes, it is worth a bit more attention. First, look at the current induced between planes when a single charge moves between them, as sketched in Fig. 3-4a. An elemental picture of current flow to

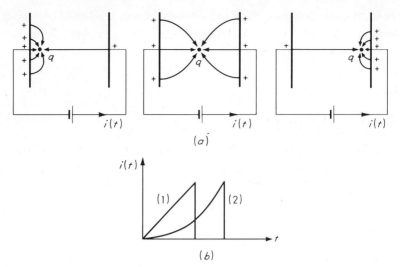

Fig. 3-4 (a) Various positions of charge moving from anode to cathode, showing redistribution of induced charges on electrodes; (b) induced current at anode. (1) Temperature-limited emission; (2) space-charged-limited emission.

the anode might lead us to believe that current flows only when the charge actually arrives at the anode. However, attention to the induced charges as indicated in the sequence of Fig. 3-4a shows that current flows in the external circuit during all the time that the charge is moving between planes; when the charge actually arrives at the anode, the negative of its charge has been transported so that current falls to zero at the instant the electron is collected (neglecting complications from secondary electrons). The picture of current flow in the external circuit might then look something like the sketch of Fig. 3-4b.

The expression for current induced in the above example is

$$i(t) = \frac{qv(t)}{d} \tag{7}$$

where d = the diode spacing
q = the charge
$v(t)$ = the velocity of the charge

This may be derived from the method of images, from Gauss' law, from Green's reciprocation theorem, from the analytic form of Green's theorem, or from energy considerations. We shall use the energy method, which is as simple as any. If the voltage applied across the planar diode in Fig. 3-4a has the constant value V_0, the instantaneous power input to the battery is

$$P(t) = V_0 i(t) \tag{8}$$

This power must equal the rate of change of the charge's kinetic energy produced by the acceleration of the charge through the mechanism of the electric field; thus

$$P(t) = \frac{d}{dt}\frac{mv^2}{2} = mv\frac{dv}{dt} = v\frac{qV_0}{d} \tag{9}$$

Equating (8) and (9), we at once get (7).

Shockley and Ramo each gave general forms of the above equation applicable to electrodes of any shape but restricted to electrode spacings small compared with the wavelength. For a multielectrode system, the current to the nth electrode induced by a charge q moving with vector velocity **v** in the vicinity may be written

$$i(t) = q\mathbf{v} \cdot \mathbf{E'} \tag{10}$$

where **E'** is a vector electric field computed at the charge position as follows: The charge is removed, the nth electrode is held 1 volt above ground, and all other electrodes are grounded. Note that if the charge velocity **v** is perpendicular to **E'**, no current is induced on the nth electrode, by (10).

It is clear that if electrons take an appreciable part of a cycle to travel between electrodes, the induced current will flow all of that time and tube performance must take this into account.

EXERCISES

3-3 Derive the generalized Eq. (10) by an energy argument similar to that used for the parallel-plane diode of this section.

3-4 Derive the expression for current induced to the anode of a parallel-plane diode when the current flow is a planar slab of charge q coul/m², moving with velocity v in the x direction. Comment on the applicability of this result to the case of the point charge by considering the principle of superposition.

3-5 An electron moves radially from the inner cylindrical electrode of a coaxial diode toward the outer cylindrical anode which is at potential V_0 with respect to the inner.

Find the value of current induced in the anode when the charge is at radius r between electrodes.

3-6 In a static or low-frequency problem, we consider the current flow to the grid lead as given by the actual current collected by the grid minus grid emission current, if any. Reconcile this result with the point of view given in this section that current flowing to the electrode will be given by the induced effects from all electrons in the space.

3-7 Show that Eq. (7) does follow from the more general Eq. (10).

3-4. TRANSIT-TIME EFFECTS IN SPACE-CHARGE-CONTROLLED TUBES

Both induced-current and charge-acceleration effects cause the characteristics of space-charge–controlled tubes (diodes, triodes, etc.) to vary from the static characteristics when the transit time between any two electrodes becomes an appreciable fraction of a radio-frequency (RF) period. Some of these effects are discussed qualitatively in the following paragraphs, and one important effect is treated in more detail in the next section.

As the transit time τ from cathode to anode in a space-charge–limited diode becomes an increasingly large fraction of the RF period, the induced currents due to charges which leave the cathode at different instants in the RF period do not add with the same phase, and as a consequence, both the small-signal ac conductance and susceptance are functions of the transit angle $\omega\tau$, as shown in Fig. 3-5.† Note that the small-signal conductance actually becomes negative when the transit angle exceeds 2π by a small amount. If a load conductance of smaller magni-

† The analysis leading to Fig. 3-5 may be found in Ref. 3.

Fig. 3-5 Space-charge–limited diode capacitance and conductance as a function of transit angle.

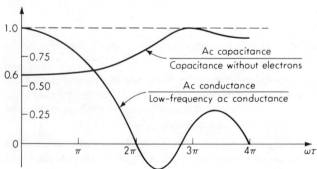

microwave communications

tude than the diode negative conductance is placed across the diode, oscillations are possible. Figure 3-5 shows that the ac capacitance is also a function of the transit angle. Since the transit time of the electrons across the diode is a function of the anode voltage, a change in anode voltage can change the capacitance of the diode at a fixed frequency. If the diode is part of a tuned circuit, this change in capacitance can detune the circuit.

The induced currents which flow in a negative-grid triode because of the passage of an electron from cathode to plate are shown in Fig. 3-6. At low frequencies the cathode-grid transit time τ_g is such a small portion of a cycle that the positive and negative portions of the grid current cancel each other. At higher frequencies, with τ_g an appreciable part of a cycle, the grid voltage may change enough between the positive and negative pulses of grid current shown to cause appreciable change in the tube operation. This physical picture will be assumed in the following discussion of transit-time effects in multielement electron tubes. Note that the net charge delivered to the grid must be zero and that the net charge delivered to the plate is equal to the net charge which *left* the cathode. Note also

Fig. 3-6 Induced currents on the electrodes of a triode due to a single electron moving from cathode to anode.

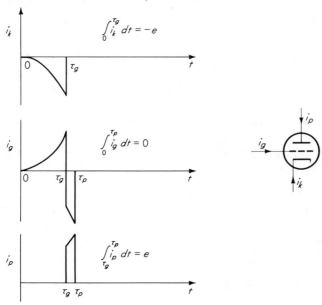

microwave amplifiers aad oscillators

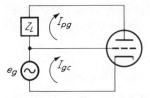

Fig. 3-7 Common-grid connection.

that the graphs of Fig. 3-6 indicate that at no instant of time t does cathode-plate current flow but that a cathode-grid current flows for $0 < t < \tau_g$ and then a grid-plate current flows for $\tau_g < t < \tau_p$. This picture is somewhat different from the low-frequency picture obtained in elementary considerations.

The input admittance of a grounded-grid triode (Fig. 3-7) is substantially the same as the diode admittance given by Fig. 3-5, because the grounded grid effectively isolates the cathode and plate circuits. Thus input loading decreases with increasing transit angle and may even become negative, causing oscillations under some conditions. The capacitance variation may cause detuning, and is one source of oscillator frequency change with bias voltage.

The input admittance of a triode in the common-cathode connection is different from the common-grid connection because the induced current in the grid-plate space flows through the input generator (Fig. 3-8) (neglecting the fraction which flows through the grid-plate capacity). For this connection, the input loading due to induced currents is very low at small transit angles, increases as the square of the transit angle for small angles, but eventually oscillates in magnitude and sign as in the common-grid connection case. In the next section we shall show that the input conductance in this connection for small transit angles is

$$g_{\text{ind}} = Kg_m\omega^2\tau^2 \quad \text{for } \omega\tau \ll 1 \tag{11}$$

Fig. 3-8 Common-cathode connection.

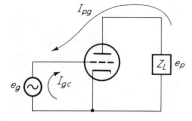

The self-capacitance and loading conductance of the grid-plate region of a triode or tetrode also change with transit angle because of the effects on induced currents, but this effect is generally of less importance than in the lower-velocity grid-cathode region.

The current induced in the grid-plate space resulting from an alternating voltage applied across the grid-cathode space suffers a phase lag as a consequence of the finite transit time, and is also decreased in magnitude. The transconductance thus becomes complex and is more properly called a *transadmittance*. Since the phase angle of the transadmittance is sensitive to voltage changes, plate modulation of the tube may produce accompanying phase- or frequency-modulation effects when the transit angle is large. The decreased magnitude of transadmittance means decreased gain, but fortunately the magnitude is more sensitive to grid-plate transit angle, which is generally small owing to the relatively higher grid-plate voltages.

Efficiency may decrease drastically in a class C amplifier operating under large-transit-angle conditions since the phase lag and spread of phases will make it difficult or impossible to collect all electrons at the anode with the desired low kinetic energies. Some electrons may also be returned to the cathode if they do not pass through the grid plane in time. The resulting cathode back bombardment causes power loss and cathode damage.

Many other complications arising from secondary emission, potential minimum effects near the cathode, and tube geometry when transit times are important cause the curves on actual tubes to be different from those sketched for the ideal planar tubes. However, a good understanding of the effects stated above will usually form a basis for analyzing the first-order phenomenon.

The transit time in a space-charge–limited diode with cathode-anode spacing of d and applied voltage V_0 is

$$\tau = \frac{3d}{(2\eta V_0)^{\frac{1}{2}}} \quad \text{sec} \tag{12}$$

where η is the ratio of electronic charge to mass (1.76×10^{11} coul/kg). In using (12) for the cathode-grid space of a triode, the grid potential should be taken as the equivalent grid-plane potential ($E_c + E_b/\mu$), where μ is the amplification factor. Alternatively, the grid potential may be more convenient to estimate from the current density as related by the Child-Langmuir relationship,

$$i = 2.33 \times 10^{-6} \frac{V_0^{3/2}}{d^2} \quad \text{amp/m}^2 \tag{13}$$

EXERCISES

3-8 At the onset of transit-time effects, assume that the first important effect is a phase change in the induced currents with respect to the ac grid-cathode voltage. Thus

$$I_{gc} \approx g_m E_{gc} e^{+i\frac{1}{10}\omega\tau} \qquad I_{pg} \approx g_m E_{gc} e^{-i\frac{1}{3}\frac{1}{6}\omega\tau}$$

Derive (11) for $\omega\tau \ll 1$ and give the constant K for this case.

3-9 Derive the transit-time expression (12).

3-10 A tetrode has cathode-grid spacing of 0.010 in., grid-screen spacing of 0.020 in., and screen-plate spacing of 0.020 in. Current density is 1,000 amp/m² under space-charge–limited conditions, screen potential is 200 volts, and plate potential is 200 volts. Estimate the transit time in each space and the frequency at which performance might be expected to change as a consequence of transit-time effects.

3-11 A 416A triode has a cathode-grid spacing of 0.001 in. Current density is 2,000 amp/m² under transit-time–limited conditions. Estimate the cathode-grid transit angle at 4,000 Mc under ideal assumptions, and state why these assumptions might be poor for such close electrode spacings.

Triode Input Admittance Variation with Small Transit Angles

Due to transit-time effects the phase difference between the induced current in the grid and the impressed grid voltage causes the grid loading (input conductance) in a common-cathode triode amplifier to increase as the square of the signal frequency. If q_{ind} is the induced charge on the grid due to the electrons flowing from cathode to plate in the negative-grid tube, then the total charge on the grid due to the impressed voltage e_g may be written as

$$q_g = q_{\text{ind}} + C_{gk} e_g + C_{gp}(e_g - e_p) \tag{14}$$

The subscripts used in (1) imply that we are considering only small-signal alternating quantities. The induced charge q_{ind} is certainly proportional to the charge in transit between the cathode and plate at any one instant in time. If we write the impressed grid voltage as

$$e_g = E_g e^{j\omega t} \tag{15}$$

and the transit time for a single electron from cathode to grid under the dc bias conditions of the tube is τ_g, then the induced charge is of the form

$$q_{\text{ind}} = K\tau_g g_m E_g e^{j\omega(t-\tau_g)} \tag{16}$$

Here $g_m E_g \tau_g$ is proportional to the charge which is emitted from the cathode at time t and passes the grid τ_g sec later, and K is a proportionality constant not greatly different from unity. (Exercise 3-12 illustrates in a different manner that the above expression is correct.) The current induced in the grid by the summation of all charges moving through the triode is the time derivative of q_{ind}:

$$i_{\text{ind}} = jK\omega\tau_g g_m E_g e^{j\omega(t-\tau_g)} \tag{17}$$

This current variation is drawn in Fig. 3-9. As the induced charge is nearly in phase with the grid-cathode voltage e_g for small transit angles, the induced current is nearly 90° out of phase with e_g. The zero in grid current occurs when the electron which is emitted when e_g is maximum just passes the grid. (See Exercise 3-13.) When the transit angle $\omega\tau_g$ is small compared with unity, (17) can be simplified to

$$i_{\text{ind}} \approx jK\omega\tau_g g_m(1 - j\omega\tau_g + \cdots)E_g e^{j\omega t} \tag{18}$$

or

$$i_{\text{ind}} = Y e_g \approx (\omega^2 \tau_g^2 g_m k + j\omega\tau_g g_m k)e_g \tag{19}$$

The term in parentheses in (19) is the input admittance due to transit-

Fig. 3-9 Graph showing current due to one electron emitted (when e_g is maximum) and the total induced grid current i_{ind}.

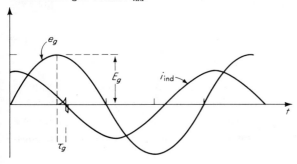

time effects for a grounded-cathode triode amplifier and small transit angles. Note that there is an imaginary part which represents a capacitive susceptance. This capacitance is independent of frequency, but depends linearly on the transit time. There is also a conductance which increases as the square of the transit angle and which loads the input circuit of the triode at high enough frequencies. The capacitance exists whenever charge flows, and is the reason the hot capacitance of a tube differs from the cold capacitance. It can detune the input circuit at high frequencies.

EXERCISES

3-12 Show that the magnitude of the frequency spectrum of any doublet pulse of the form shown in Fig. 3-10 is $|F(j\omega)| \propto \omega\tau$, for $\omega\tau \ll \pi$. Use this result to justify (17).

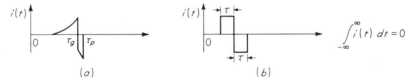

Fig. 3-10 Current doublet pulses.

3-13 Assume that the grid current due to a single electron passing the grid at time τ is given by Fig. 3-10b. If the total grid current is the sum of millions of such pulses and the maximum sinusoidal grid voltage occurs at time $t = 0$, then $e_g = E_g \cos \omega t$ and $i_g = g_m E_g [\cos \omega(t - \tau) - \cos \omega(t - 2\tau)]$. Show that this equation for i_g will yield the same form for the input admittance under the limiting assumption $\omega\tau \ll 1$ as (19).

3-14 Determine the total charge between cathode and anode of a space-charge-limited triode, and show that for constant applied voltages and small signals the form of this "charge in transit" is given by (16). Discuss why this charge is *not* the same as the induced charge on the grid.

3-5. KLYSTRONS

In the late 1930s it was becoming apparent that the high-frequency performance of gridded vacuum-tube amplifiers was limited by parasitic circuit elements and by transit-time effects. The former limitation was ultimately overcome by making the tube elements part of a resonant cavity; the transit-time limitation seemed fundamental. In 1935 A. A.

Heil and O. Heil[4] published a paper describing the generation of high-frequency voltages using transit-time effects together with lumped tuned circuits, but their paper received little attention. In 1939 W. C. Hahn and G. F. Metcalf[5] published a paper describing many possible velocity-modulated tubes and giving the basic theory for their analysis. Four months later R. H. Varian and S. F. Varian[6] described a two-cavity klystron amplifier and oscillator which employed resonant cavities and therefore was much more efficient in modulating the electron beam with a given input power. They were able to achieve appreciable electronic gain and power generation at centimeter wavelengths.

Like many other components of communications systems, the klystron was developed extensively during World War II to meet the urgent need for reliable microwave oscillators and amplifiers, and since the war, this development has continued. Today the klystron is used primarily as the power-output tube for UHF television stations, for various microwave communication systems (both civilian and military), and for linear accelerators designed to increase man's knowledge of the fundamental structure of matter.

A schematic diagram of a klystron amplifier is shown in Fig. 3-11. The space-charge–limited current from the thermionic cathode is accelerated through the anode hole in a parallel electron beam. The shape of the

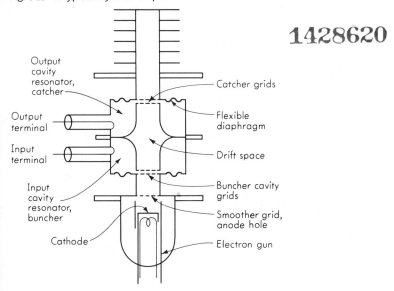

Fig. 3-11 A typical klystron amplifier.

beam-forming electrode is designed to produce a potential between cathode and anode which varies as $z^{\frac{4}{3}}$, that is, as the potential between cathode and anode in an infinite parallel-plane diode. The cathode, beam-forming electrode, and anode, taken together, are called a Pierce electron gun. Microwave electron tubes often employ long electron beams; the formation and focusing of these beams, although an important subject for the microwave tube designer, are beyond the scope of this text. The interested reader is referred to books by Spangenberg[7] and Pierce.[8]

In Fig. 3-11, the electrons enter the input cavity with uniform velocity u_0; while passing through the cavity they are accelerated by the field which exists between the cavity grids, either gaining or losing energy depending upon the phase of the voltage across the gap as they pass through it. The electrons then drift in the field-free region separating the two cavity gaps. The electrons accelerated by the input cavity field travel slightly faster than u_0 in the drift space and slowly catch up with the unaccelerated electrons; similarly, electrons decelerated by the input cavity field are gradually overtaken by the unaccelerated electrons, and an electron bunch forms about these unaccelerated electrons, as is shown in the distance-time plot (or Applegate diagram) of Fig. 3-12. If the

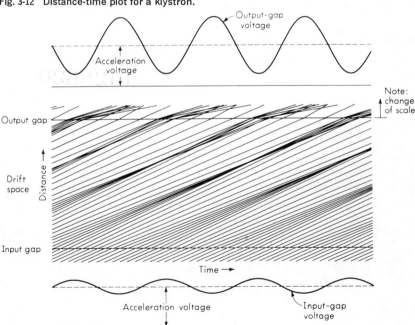

Fig. 3-12 Distance-time plot for a klystron.

average charge density of the beam is low enough, electrons will actually pass each other, as is shown in the distance-time plot. If the charge density of the beam is high, as is often the case in high-power tubes, the space-charge forces between electrons must be taken into account; these forces modify the trajectories shown in Fig. 3-12 and also cause the beam to spread transversely.

In the sections which follow, gridded klystron gaps are analyzed. Actually, klystron amplifiers with two or more cavities usually employ gridless cavities to avoid power dissipation caused by electrons striking the grids. The principle of operation of both gridded and gridless gaps is the same, and the mathematical analysis of the former is much easier; hence only gridded gaps are considered below.

There are fundamental limitations in the power output which can be expected from klystrons and most other microwave tubes. Present-day klystron amplifiers are capable of delivering pulse power output in the 1- to 30-Mw range and average power outputs up to several hundred kilowatts at the lower microwave frequencies. The beam voltage of such klystrons is of the order of 100 kv, and as the efficiency of these devices is of the order of 30 to 50 percent, the peak current must be of the order of a few hundred amperes. Furthermore, a 10-Mw klystron with an efficiency of 40 percent must dissipate some 6 Mw of peak power inside the tube itself. This is a tremendous amount of power to be dissipated inside a vacuum tube, and very elaborate precautions are taken so that these tubes can operate over long periods of time. They are generally constructed with a built-in vacuum pump which operates for the life of the tube. Special metals and ceramics are used which when sealed together form a rugged ultrahigh-vacuum envelope. These tubes are large, of the order of 4 to 10 ft in length, and generally the beam is focused by an axial magnetic field in order to overcome space-charge spreading. The power density in the electron beam runs many megawatts per square centimeter, and consequently the beam must be accurately focused through the device. If the beam strikes surfaces not designed to dissipate its high power density, it quickly destroys them.

Once the large microwave power is generated inside the output cavity, it still must be coupled through a vacuum seal into a waveguide for transmission to the load. The vacuum seals are generally thin dielectric windows across the waveguide; the very high electric fields which pass through these dielectric windows can cause breakdown, and sometimes electron emission is observed. The reliability and capability of present-day high-power klystrons are indeed a tribute to the many people who have worked both to understand klystron theory and to overcome the technological problems connected with manufacturing. The same sort of tribute also applies to those persons who have through their research and development efforts made possible the other components discussed throughout this text.

3-6. VELOCITY MODULATION BY A GRIDDED CAVITY GAP

The velocity of the electrons passing between the input cavity grids is modified by the radio-frequency electric field between the grids. The effect of the beam space charge on the electric field between the grids is neglected (generally a very good assumption), and therefore the motion of the beam through the cavity gap may be determined by finding the corresponding motion of a single electron. Throughout the following analysis, e will represent the *magnitude* of the electronic charge ($e = 1.6 \times 10^{-19}$ coul), and $e/m = \eta$ will represent the electronic charge-to-mass ratio ($\eta = 1.76 \times 10^{11}$ coul/kg).

In Fig. 3-13, the electron beam enters the input cavity from the left with a uniform velocity $u_0 = (2\eta V_0)^{\frac{1}{2}}$ at time t_0. If the second grid is positive with respect to the first grid, the gap voltage $V = V_1 \sin \omega t$ is taken to be positive and the electric field across the gap, which is determined by V and the gap spacing d, accelerates the electron beam. The exit velocity of electrons which entered the gap at time t_0 will now be

Fig. 3-13 Schematic diagram of a gridded klystron cavity.

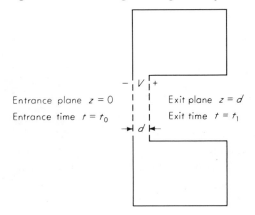

found. Using the force equation, the acceleration of the electron is

$$\frac{d^2z}{dt^2} = \ddot{z} = -\eta E_{\text{gap}} = +\eta \frac{V_1}{d} \sin \omega t \qquad (20)$$

Integrating the acceleration with respect to time from t_0 to t,

$$\int_{t_0}^{t} \ddot{z}\, dt = \dot{z}(t) - \dot{z}(t_0) = -\eta \frac{V_1}{\omega d}(\cos \omega t - \cos \omega t_0) \qquad (21)$$

An additional integration yields

$$z(t) - z(t_0) = \dot{z}(t_0)(t - t_0) + \eta \frac{V_1}{\omega d}(t - t_0)\cos \omega t_0$$

$$- \eta \frac{V_1}{\omega^2 d}(\sin \omega t - \sin \omega t_0) \qquad (22)$$

Using the boundary condition that $z(t_0) = 0$ and $z(t_1) = d$ and the initial condition that $\dot{z}(t_0) = u_0$, (22) can be evaluated at $t = t_1$, the moment the electron exits from the gap, giving

$$d = u_0(t_1 - t_0) + \eta \frac{V_1}{\omega d}(t_1 - t_0)\cos \omega t_0 - \eta \frac{V_1}{\omega^2 d}(\sin \omega t_1 - \sin \omega t_0) \qquad (23)$$

Only a moderate amount of algebra is necessary to show that if $V_1 \ll V_0$, the second and third terms of the above expression are very nearly zero and (23) reduces to

$$d \approx u_0(t_1 - t_0) \qquad (23a)$$

Solving (23a) for t_1 and substituting into (21), we find the exit velocity from the gap to be

$$v(t_1) = u_0 - \eta \frac{V_1}{\omega d}\left[\cos \omega \left(t_0 + \frac{d}{u_0}\right) - \cos \omega t_0\right] \qquad (24)$$

The quantity $\omega d/u_0$ is recognized as the gap transit angle when $V_1 = 0$ and $\theta_g = \omega(t_1 - t_0)$; with the help of a trigonometric identity and the relationship between u_0 and V_0, we can rewrite (24) as

$$v(t_1) = u_0\left[1 + \frac{V_1}{2V_0}\frac{\sin(\theta_g/2)}{\theta_g/2}\sin\left(\omega t_0 + \frac{\theta_g}{2}\right)\right] \qquad (24a)$$

This expression is the small-signal electron velocity leaving the klystron gap. The factor $[\sin(\theta_g/2)]/(\theta_g/2) \equiv \beta$ is called the beam-coupling coefficient and is best understood by considering Fig. 3-14. For efficient coupling, θ_g should be kept small. Increasing θ_g decreases the coupling between the beam and the cavity, i.e., decreases the velocity modulation of the beam for a given cavity voltage.

The distance-time diagram of Fig. 3-12 is obtained by taking $v(t_1)$ as the electron velocity in a field-free region (drift tube); thus velocity modulation leads directly to the electron bunching, or current modulation, of the klystron beam.

In arriving at Eq. (23a), we have retained terms containing one small, or first-order, quantity such as V_1/V_0, while terms containing the product of two first-order terms (and thus called second-order terms) have been neglected. This procedure is extensively used and yields quite good results for first-order quantities, such as velocity, current, or voltage. However, second-order terms are *not* correctly calculated from the first-order theory, and they often are important. For example, the time-average power absorbed by the beam in passing through the input cavity might be calculated by determining the time-average increase in kinetic energy of the electron beam, using Eq. (24a). This power would be incorrect. Beck[9] has carried through an analysis correct to second order which correctly gives the power absorbed by the beam in passing through the input cavity. This power is

$$P = \frac{V_1^2 I_0}{2V_0}\left(\frac{1 - \cos\theta_g}{\theta_g^2} - \frac{\sin\theta_g}{2\theta_g}\right) \tag{25}$$

Since this power is proportional to the square of the peak voltage across

Fig. 3-14 Beam-coupling coefficient β versus gap transit angle θ_g.

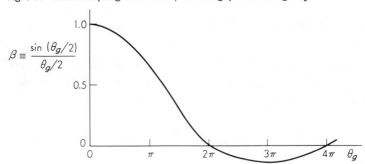

microwave communications

the cavity gap, the beam loading can be represented as a conductance,

$$g_b = \frac{2P}{V_1^2} = \frac{I_0}{V_0}\left(\frac{1-\cos\theta_g}{\theta_g^2} - \frac{\sin\theta_g}{2\theta_g}\right) \tag{26}$$

When the cavity is represented by a resonant circuit, the loss in the cavity determines the cavity Q in the absence of the beam, but in the presence of the electron beam, the beam-loading conductance given by (26) must also be included. Beck also shows that a susceptance appears across the cavity because of the beam, and thus the beam may slightly detune the cavity. The value of the susceptance is

$$b_b = \frac{I_0}{V_0}\left(\frac{\sin\theta_g}{\theta_g^2} - \frac{1+\cos\theta_g}{2\theta_g}\right) \tag{27}$$

EXERCISES

3-15 Show analytically that (23a) follows from (23) if $V_1 \ll V_0$.

3-16 Show that (24a) follows from (24).

3-7. THE TWO-CAVITY KLYSTRON

In this section we shall determine how the velocity modulation on the electron beam produced by the first cavity of a two-cavity klystron is transformed into current modulation in the drift space. The current-modulated beam induces currents (and fields) in the second cavity, producing an output signal which is an amplified version of the input. If the input signal is too high, the output signal will not be a linear version of the input; an analytic expression for the nonlinear relationship between input and output will be obtained. The coordinates and their associated times for a given electron are shown in Fig. 3-15.

From (24a) of the previous section, it is convenient to write the velocity of the electrons just entering the drift space between input and output cavity gaps in terms of t_1, the time the electrons enter the drift space. Thus

$$v(t_1) = u_0\left[1 + \frac{V_1}{2V_0}\beta\sin\left(\omega t_1 - \frac{\theta_g}{2}\right)\right] \tag{28}$$

Since no fields exist in this section, the time an electron entering at t_1

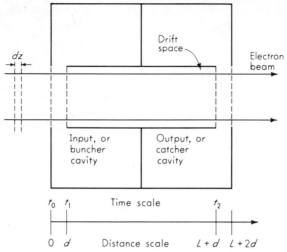

Fig. 3-15 Schematic diagram of a two-cavity klystron.

reaches the output cavity gap is given by

$$t_2 - t_1 = \frac{L}{v(t_1)} = \frac{L}{u_0}\left[1 - \frac{V_1}{2V_0}\beta \sin\left(\omega t_1 - \frac{\theta_g}{2}\right)\right] \quad (29)$$

where the binomial expansion of $(1 + y)^{-1}$ has been used, assuming $y = V_1\beta/2V_0 \ll 1$. This expression will be used shortly.

First we must quantitatively determine the beam current as a function of position along the drift tube. This is accomplished by invoking the conservation of charge.

Consider an infinitesimal length of the electron beam dz just before the input cavity. If the charge per unit length of the beam is ρ, the charge dQ enclosed by two planes perpendicular to the beam axis and positioned at z and $z + dz$ is $dQ = \rho\,dz$. If these planes move with the electron beam, always keeping the same total charge dQ between them, the first would pass a fixed point z_0 at time z_0 and the second at time $t_0 + dt_0$, where $dt_0 = dz/v(t_0)$. Since the current in the beam is $I = \rho v$, the charge between the two planes can be written as

$$dQ = \rho(t)\,dz(t) = \rho(t)v(t)\,dt = I(t)\,dt \quad (30)$$

At the input grid of the first gap, I has the dc value I_0, and t is by definition t_0. After the velocity modulation of the input cavity, ρ, v, and dz will all change, and the current at the output cavity I_2 at time t_2 can be found from the relationship

$$I_2|dt_2| = I_0|dt_0| \tag{31}$$

The absolute-value signs are required by physical considerations which will be discussed below. Beck[10] has derived this same equation more rigorously. Rewriting (29) in terms of the initial time t_0, we have

$$t_2 = t_0 + \frac{d}{u_0} + \frac{L}{u_0}\left[1 - \frac{V_1}{2V_0}\beta \sin\left(\omega t_0 + \frac{\theta_g}{2}\right)\right] \tag{32}$$

and then differentiating yields

$$dt_2 = dt_0\left[1 - \frac{\omega L}{u_0}\frac{V_1}{2V_0}\beta \cos\left(\omega t_0 + \frac{\theta_g}{2}\right)\right] \tag{33}$$

The current at the entrance to the second cavity is then

$$I(t_2) = \frac{I_0}{1 - (\omega L/u_0)(V_1/2V_0)\beta \cos(\omega t_0 + \theta_g/2)} \tag{34}$$

Note that the time an unmodulated electron would require to travel between input and output cavity is L/u_0, so the dc transit angle is merely

$$\theta_0 = \frac{\omega L}{u_0} \tag{35}$$

The bunching parameter of a klystron is defined as

$$X \equiv \frac{\theta_0 V_1 \beta}{2V_0} \tag{36}$$

Assuming that $X \ll 1$ (which is the most serious restriction so far to our analysis, since θ_0 may be several radians), the current entering the second cavity is

$$I(t_2) = I_0\left[1 + X \cos\left(\omega t_0 + \frac{\theta_g}{2}\right)\right] \tag{37}$$

Thus for small signals, that is, $X \ll 1$, the beam current at the second cavity consists of a dc current plus a component at the input frequency.

In terms of t_2,

$$I(t_2) = I_0 \left[1 + X \cos\left(\omega t_2 - \theta_0 - \frac{\theta_g}{2}\right) \right] \tag{37a}$$

The current modulation on the beam will induce a current in the output cavity which is given by qv/d. The alternating charge q in transit is

$$q = \int_{t_2 - \frac{d}{u_0}}^{t_2} I(t_2)\, dt_2 = \frac{2I_0 X \sin\frac{\theta_g}{2}}{\omega} \cos(\omega t_2 - \theta_0 - \theta_g) \tag{38}$$

Therefore, the alternating current induced in the output cavity is

$$I_{\text{ind}} = \frac{qu_0}{d} = I_0 X \beta \cos(\omega t_2 - \theta_0 - \theta_g) \tag{39}$$

where velocity modulation on the beam has been ignored. An equivalent small-signal transconductance of the two-cavity klystron may be defined as

$$g_m = \frac{I_{\text{ind}}}{V_1} = \frac{I_0}{V_0} \frac{\beta^2 \theta_0}{2} \tag{40}$$

The klystron can now be replaced by an equivalent current generator, as shown in Fig. 3-16, where g_L is the equivalent load conductance shunted across the cavity and g represents the sum of the inherent cavity loss and the beam-loading conductance. At resonance, the power output is

$$P_{\text{out}} = \frac{1}{2} \frac{g_m^2 V_1^2}{(g + g_L)^2} g_L \tag{41}$$

Proceeding now to the less restrictive case where X may be appreciable compared with unity, let us determine graphically the current as a function of time at the output cavity. It is convenient to first draw a

Fig. 3-16 Klystron equivalent circuit.

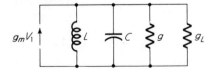

graph of departure angle from the input cavity's center plane versus the arrival angle at the output cavity; this is the top graph of Fig. 3-17. (The departure and arrival angles are directly proportional to the departure and arrival times of an electron.)

The current arriving at the output cavity is now easily determined graphically by using (31). The coordinates of the departure and arrival angles were chosen as the most natural ones, since (32) when rewritten is

$$\omega t_2 - \left(\theta_0 + \frac{\theta_g}{2}\right) = \left(\omega t_0 + \frac{\theta_g}{2}\right) - X \sin\left(\omega t_0 + \frac{\theta_g}{2}\right) \qquad (42)$$

Note that infinities of current result when the ratio $dt_2/dt_0 = 0$. These infinities are a result of our model, which excludes the space-charge forces (mutual repulsion of electrons). The infinities occur when two planes containing a finite charge between them move together until they coincide. A negative value of dt_2/dt_0 indicates that certain electrons overtake and pass others between the input and output cavities. A negative value of

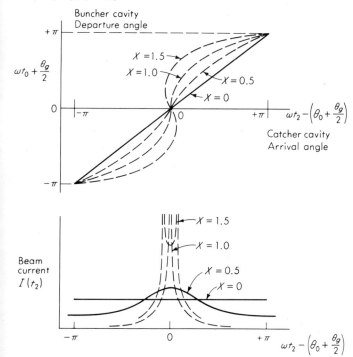

Fig. 3-17 Graphs illustrating method of determining beam-current modulation at the catcher (output) cavity.

this ratio would indicate a negative current if the absolute-value signs were omitted in (31); thus these signs are necessary to yield the correct physical result.

We shall proceed with the analysis and continue to neglect space charge since the analytic results have been found in reasonable agreement with experiment in many cases. Note that the beam current at the second cavity is a periodic waveform of period $2\pi/\omega$. We therefore can expand it in a Fourier series and find the harmonic content. Thus

$$I(t_2) = I_0 + \sum_{n=1}^{\infty} I_n \cos n\left(\omega t_2 - \theta_0 - \frac{\theta_g}{2}\right) \tag{43}$$

where

$$I_n = \frac{1}{\pi} \int_{-\pi}^{\pi} I(t_2) \cos n\left(\omega t_2 - \theta_0 - \frac{\theta_g}{2}\right) d(\omega t_2) \tag{44}$$

It is necessary to invoke the conservation of charge, (31) and (42), yielding

$$I_n = \frac{1}{\pi} \int_{-\pi}^{\pi} I_0 \cos\left[n\omega t_0 + \frac{n\theta_g}{2} - nX \sin\left(\omega t_0 + \frac{\theta_g}{2}\right)\right] d(\omega t_0) \tag{45}$$

or, rewriting in terms of a dummy variable y,

$$I_n = \frac{1}{\pi} \int_{-\pi}^{\pi} I_0 \cos(ny - nX \sin y) \, dy \tag{45a}$$

Evaluating (45a), we find

$$I_n = 2I_0 J_n(nX) \tag{46}$$

where $J_n(nX)$ is a Bessel function of the first kind, of order n and argument nX.[11] Therefore the beam current at the catcher cavity is

$$I(t_2) = I_0 + \sum_{n=1}^{\infty} 2I_0 J_n(nX) \cos n\left(\omega t_2 - \theta_0 - \frac{\theta_g}{2}\right) \tag{47}$$

which mathematically is known as a Fourier-Bessel series. The values of $J_n(z)$ are plotted versus z for $n = 0, 1,$ and 2 in Fig. 3-18; the magnitude of

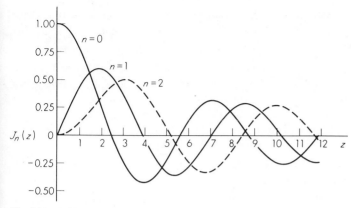

Fig. 3-18 $J_n(z)$ versus z.

the fundamental component of beam current at the catcher cavity is

$$I_1 = 2I_0 J_1(X) \tag{48}$$

Equation (48) states the peak fundamental component of beam current in terms of the peak voltage across the input cavity gap, since X, the bunching parameter, is directly proportional to V_1. As shown above, the fundamental component of current induced in output cavity by the bunched beam is βI_1. An equivalent nonlinear transconductance of the klystron can be defined as

$$g_m = \frac{\beta I_1}{V_1} = \frac{\beta^2 I_0 \theta_0}{2V_0} \left[\frac{2J_1(X)}{X} \right] \tag{49}$$

The bracketed term in (49) is called the saturation factor and is plotted in Fig. 3-19. As the input voltage V_1 increases, X increases for a given

Fig. 3-19 Klystron saturation factor.

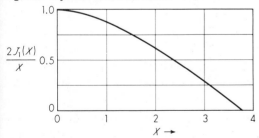

microwave amplifiers and oscillators

average beam voltage and transit angle and the saturation factor decreases, causing a decrease in the transconductance. The output power reaches a maximum when $J_1(X)$ is maximum, which occurs at $X = 1.84$.

EXERCISES

3-17 Determine the current induced in the output cavity by the velocity modulation on the beam, and thus determine a condition for (39) to be valid.

3-18 Determine the maximum available power gain of a two-cavity klystron with identical input and output cavities.

3-19 A two-cavity klystron has 1 watt of output power when the bunching parameter is unity. The input power is then doubled. Determine the new value of the output power.

3-20 The cathode voltage of a 3-Gc two-cavity klystron is modulated with a 1-kc sawtooth waveform, such that the peak-to-peak value of the sawtooth is 2 percent of the average cathode voltage V_0. Compare the output power and frequency of this klystron with those quantities when no cathode modulation is present. Assume the cathode current I_0 is constant.

3-8. THE REFLEX KLYSTRON

In the above analysis of a two-cavity klystron, the cavities were completely decoupled from each other. If a small fraction of the output power is fed back to the input cavity and if the phase of this feedback signal is adjusted to reinforce the input signal, greater gain will result through regeneration. If enough output power is fed back, oscillations will occur. However, a two-cavity klystron oscillator has certain disadvantages; to change the oscillation frequency, for example, not only must the resonant frequency of each cavity be altered but also the feedback path phase shift must be readjusted to give positive feedback.

These disadvantages are overcome by the reflex klystron, a single-cavity klystron oscillator in which the electrons themselves provide the feedback, as shown in Fig. 3-20. The repeller (or reflector) voltage is always less than the cathode voltage. Electrons are accelerated from the cathode to the cavity gap, where they are velocity-modulated by the RF voltage across the gap. These electrons then penetrate into the repeller region, which is taken here as a constant retarding-field region, where they are slowed down, reflected, and accelerated back toward the cavity gap. Bunching occurs during this transit through the retarding field,

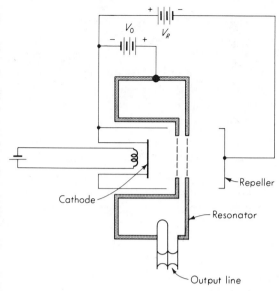

Fig. 3-20 Schematic diagram of a reflex klystron.

because electrons accelerated by the gap voltage penetrate farther into the repeller region and thus take longer to return to the gap than do electrons which passed through the gap when the RF voltage was zero. Similarly, electrons decelerated by the field do not penetrate so far into the repeller region and remain there a shorter time than unaccelerated electrons.

The distance-time diagram for the reflex klystron of Fig. 3-21 demonstrates this bunching about the unaccelerated electron which passes

Fig. 3-21 Distance-time diagram for a reflex klystron.

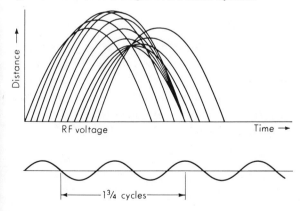

microwave amplifiers and oscillators

through the cavity gap just as the field changes from accelerating to decelerating.

If the RF electric field across the cavity gap decelerates the returning electron bunch, the net kinetic energy lost by the bunched beam will appear in the cavity as RF energy. Figure 3-21 shows that the cavity field will retard the returning electron bunch if the transit angle is $2\pi(n + \frac{3}{4})$, where the mode number n is any positive integer, including zero. If the power delivered to the cavity by the bunched beam is greater than that absorbed by the ohmic loss in the cavity and load plus the small amount of power required to velocity-modulate the beam (this latter quantity is often neglected), the electric field across the cavity will increase; i.e., oscillations will build up from the noise inherent in thermionic electron beams.

The reflex klystron has proved an inexpensive and reliable generator of milliwatts of microwave power. It is the tube most often used as the local oscillator in microwave superheterodyne receivers and microwave signal generators. The analysis below will demonstrate that it is electronically tunable over a limited frequency range, and therefore its output frequency can be easily modulated. It can be pulsed off and on quickly by pulsing either the beam voltage V_0 or the repeller voltage V_R. If there is sufficient loading of the output cavity, the reflex klystron can also be employed as a negative-resistance one-port amplifier.

The analysis of the reflex klystron is similar to the two-cavity-klystron analysis just completed. An electron entering the cavity gap from the cathode at $z = 0$ and time t_0 is assumed to have velocity $u_0 = \sqrt{2\eta V_0}$. The same electron leaves the cavity gap at $z = d$ at time t_1, with velocity

$$v(t_1) = u_0 \left[1 + \frac{\beta V_1}{2V_0} \sin\left(\omega t_1 - \frac{\theta_g}{2}\right) \right] \tag{50}$$

This expression is identical with Eq. (28) since the problems up to this point are formally identical. This electron is forced back to the cavity at position $z = d$ at time $t = t_2$ by the retarding field E, which is assumed to be constant, independent of z. All motion in a transverse direction is ignored, since all transverse electric fields are assumed to be zero.

The equation of motion in the repeller region is

$$\ddot{z} = -\eta E \tag{51}$$

This equation may be integrated, as in the two-cavity klystron case, to determine z as a function of time. The transit time $(t_2 - t_1)$ in the repeller region is found by taking $z(t_2) = z(t_1) = d$. The details of this calculation are left as an exercise; the result is

$$\omega(t_2 - t_1) = \theta_0 \left[1 + \frac{V_1\beta}{2V_0} \sin\left(\omega t_1 - \frac{\theta_g}{2}\right) \right] \tag{52}$$

where the transit angle θ_0 of an electron not accelerated by the input cavity is

$$\theta_0 = \frac{2\omega u_0}{\eta E} \tag{53}$$

The current modulation on the electron beam as it returns from the repeller region through the cavity gap is determined, as in the two-cavity-klystron case, to be

$$I(t_2) = -I_0 - \sum_{n=1}^{\infty} 2I_0 J_n(nX) \cos n(\omega t_2 - \theta_0) \tag{54}$$

where the minus signs indicate that the beam is moving in the negative z direction through the cavity gap, and θ_g has been neglected as a small quantity compared with θ_0. The fundamental component of the current induced in the cavity by the modulated beam is $I_{1\,\mathrm{ind}} = -\beta I_1$, where I_1 is the fundamental component of the beam current and β is the beam-coupling coefficient. Thus

$$I_{1\,\mathrm{ind}} = 2I_0\beta J_1(X) \cos(\omega t_2 - \theta_0) \tag{55}$$

Writing $I_{1\,\mathrm{ind}}$ in the complex form generally used for circuit analysis, we obtain

$$I_{1\,\mathrm{ind}} = 2I_0\beta J_1(X) e^{-j\theta_0} \tag{55a}$$

where $\exp(j\omega t_2)$ is understood. The voltage across the gap at time t_2 is $V_1 \sin \omega t_2$, which may also be written in complex form as

$$V = -jV_1 \tag{56}$$

with $\exp(j\omega t_2)$ understood. The ratio $I_{1\,\mathrm{ind}}/V$ is the self-admittance Y_1 of

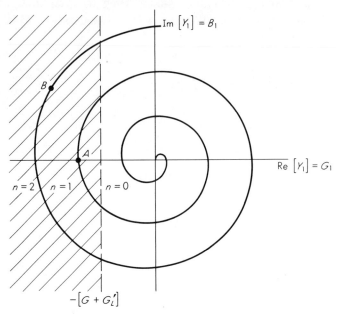

Fig. 3-22 Admittance spiral of a reflex klystron.

the reflex klystron, written as

$$Y_1 = \frac{2I_0 \beta J_1(X)}{V_1} je^{-j\theta_0} = \left[\frac{I_0}{V_0} \frac{\beta^2 \theta_0}{2} \frac{2J_1(X)}{X} \right] je^{-j\theta_0} \qquad (57)$$

The bracketed term in this equation shows that the self-admittance due to the second transit of the electron beam through the cavity gap is related primarily to the dc beam impedance I_0/V_0 and the transit angle θ_0.

The phase of Y_1 is $\pi/2$ when θ_0 is zero, and decreases linearly with θ_0; it is clear that a polar plot of Y_1 will be a spiral, as shown in Fig. 3-22, which has been drawn for $V_1 = 0$, which implies that $X = 0$. Figure 3-23 shows the equivalent circuit of the reflex klystron. In this circuit, C and L represent the energy storage elements of the cavity, G_L the load conductance transformed by the cavity coupling, and G the ohmic losses

Fig. 3-23 Equivalent circuit of a reflex klystron.

of the cavity plus the beam-loading conductance. Oscillations will occur only when Y_1 has a *negative* real part which in magnitude is greater than (or equal to) $G + G_L$; that is, only when $|-G_1| \geq G_L + G$ can enough energy be delivered from the electron beam to the cavity and external load to sustain oscillation. Thus any value of θ_0 for which the spiral lies in the shaded area of Fig. 3-22 will yield oscillation. As the spiral cuts the negative real axis at

$$\theta_0 = 2\pi(n + \tfrac{3}{4}) \tag{58}$$

the result of this analysis checks with the physical reasoning used earlier to determine the transit angle at oscillation.

The operation of the reflex klystron becomes clearer from a consideration of the admittance spiral and the saturation factor $2J_1(X)/X$, as shown in Fig. 3-19. Suppose that θ_0 is adjusted to a value of 3.5π, which corresponds to point A of Fig. 3-22. Since the negative conductance is greater in magnitude than $G_L + G$, an unstable situation exists and the RF energy in the cavity, which is proportional to V_1^2, will increase. Since V_1 is proportional to X, as V_1 becomes larger Y_1 will decrease, because Y_1 is proportional to the saturation factor. This limiting action due to the saturation factor causes the admittance spiral to shrink until point A lies at the edge of the shaded region. This is the equilibrium position; all energy delivered to the circuit by the beam is dissipated in the load and the circuit loss. The important physical point here is that the nonlinear factor in X_1, which we have called the saturation factor, limits the amplitude of oscillation. Nonlinearities limit the amplitude of any oscillation, but rarely is it possible to calculate in a relatively simple fashion *what* value the resulting oscillation amplitude will be.

If the value of θ_0 lies between 5.5π and 6π, at point B in Fig. 3-22, a similar limiting phenomenon will shrink the admittance spiral, until point B lies at the edge of the shaded region. The effect of the beam susceptance B_1 (which is capacitive in this case) is to alter the frequency of oscillation (decrease the frequency in this case). θ_0 is proportional to $fV_0^{\frac{1}{2}}/(V_R + V_0)$ since E is directly related to the repeller voltage. Since a small variation in V_R results in a small shift in oscillation frequency, frequency modulation of a reflex klystron is easily accomplished by modulating the repeller voltage. A high-impedance modulation source may be used since the repeller draws no current. Figure 3-24 shows the electronic admittance, output frequency, and output power of a

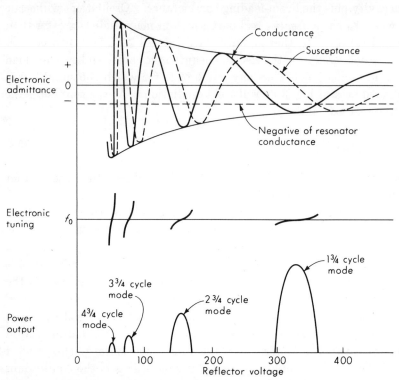

Fig. 3-24 Reflex klystron parameters versus reflector voltage.

reflex klystron as a function of repeller voltage, graphically demonstrating both the detuning and limiting discussed above.

Power and Efficiency Considerations

The power output of a reflex klystron is obtained by setting the sum of the conductances in Fig. 3-23 equal to zero and solving for the saturation factor; with the aid of Fig. 3-19 the bunching parameter X is then determined, and since I_0, V_0, and θ_0 are known in any practical case, V_1 can be determined. The power output is then

$$P = \frac{G_L V_1^2}{2} \tag{59}$$

Following the procedure of the preceding paragraph,

$$\frac{2J_1(X)}{X} \frac{\beta^2 I_0 \theta_0}{2V_0} = G + G_L \tag{60}$$

Multiplying the power output by the left-hand side of (60), dividing by the right-hand side, and using the definition of X, the power output can be put in the following useful form:

$$P = I_0 V_0 \frac{2X J_1(X)}{\theta_0} \frac{G_L}{G_L + G} \tag{61}$$

The first term ($I_0 V_0$) of the power output is the dc power input, and therefore the product of the second and third terms is the klystron efficiency. The second term is the *electronic* efficiency, and the third term is the *circuit* efficiency. The factor $X J_1(X)$ of the electronic efficiency term reaches a maximum value of 1.25 at $X = 2.408$. For any mode at maximum power output, $X = 2.408 = V_1 \theta_0 \beta / 2V$. Since θ_0 will increase for higher mode numbers, V_1 must decrease and the power output also must decrease. In practice the $n = 1$ or $n = 2$ modes have the most power output. The maximum value of $2X J_1(X)$ is about 2.5, and generally the large value of X necessary to attain this value implies that $\theta_0 \geq 3.5\pi$. The maximum electronic efficiency can therefore be estimated at $\theta_0 = 3.5\pi$ to be

$$\eta_{\max} = \frac{2.5}{3.5\pi} \approx 20\text{--}25\% \tag{62}$$

For a given value of dc power input and electronic efficiency, G_L should be large compared with G. Since G_L cannot in practice be altered without changing V_1, an optimum value of G_L will exist for any given set of dc parameters I_0, V_0, θ_0, and β. Reflex klystrons are generally designed to work into standard loads, and G_L is usually large enough so that the circuit efficiency can be considered unity.

EXERCISES

3-21 Under what circumstances will the power output from the $n = 2$ mode be greater than that from the $n = 1$ mode?

3-22 When the electron beam is off, a reflex klystron's cavity is tuned to frequency f_0 and its quality factor is measured to be Q. The beam is then turned on, and the repeller voltage is adjusted until the tube oscillates at frequency f_0. Neglecting beam loading of the cavity and assuming a constant retarding field, show that the slope of oscillation frequency versus repeller voltage evaluated at this point is

$$\frac{df}{dV_R} = \frac{\theta_0}{2Q} \frac{f_0}{V_R + V_0}$$

Is this slope greater for the $n = 1$ or $n = 2$ modes? Justify.

3-9. TRAVELING-WAVE TUBES

Microwave amplifiers considered in previous sections have utilized resonant circuits to generate large RF voltages which can efficiently interact with electrons to produce electronic gain. Amplifiers of this sort have a basic limitation, an inherent gain-bandwidth product which cannot be exceeded. Since the information capacity of the amplifier is proportional to its bandwidth, the electronic gain must decrease as the information capacity increases. The traveling-wave tube (TWT) avoids an inherent gain-bandwidth limitation by using a distributed slow-wave circuit instead of resonant circuits. An electromagnetic wave propagates along this distributed circuit with a phase velocity in the axial direction small compared with the velocity of light. An axial electron beam with a velocity approximately the same as the wave velocity interacts with the wave; since the electrons in the beam remain at the same relative phase of the wave over a long distance, the cumulative velocity modulation which results can be quite appreciable and it produces a current modulation on the beam. This current modulation, in turn, induces additional currents on the circuit, and the electric fields associated with this induced wave tend to retard the electron bunches, which give up their kinetic energy to the wave. This process is cumulative and results in an exponential growth of the wave along the circuit as long as the beam velocity is not decreased below a value necessary to maintain synchronism between beam and wave. The bunching process for TWT interaction is illustrated schematically in Fig. 3-25. Maximum gain occurs when the beam velocity is slightly greater than the above wave phase velocity, which places the electron bunches in a decreasing field; i.e., they move forward slightly in Fig. 3-25. The gain of the traveling-wave tube can be quite high; a gain of 35 db is usual, and gains of over 60 db have been reported for commercially available tubes.

Traveling-wave-tube bandwidth is limited by the frequency range over which the slow-wave circuit will propagate a constant-velocity wave and by the impedance match between the slow-wave circuit and the transmission line (or waveguide) connecting it to other components in the system. Power reflected from both the output match (or a load not properly matched to the transmission line) and the input match will be amplified by the TWT, resulting in regeneration. If the loop gain equals unity, oscillations result. To reduce the loop gain, distributed attenuation, which greatly attenuates the backward-traveling wave, is placed near the

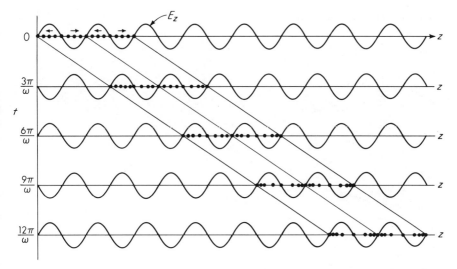

Fig. 3-25 Schematic diagram illustrating TWT bunching; the axial electric field is plotted versus distance at several different times and the effect of the wave on eight electrons contained in one period of the wave is shown. The arrows indicate the direction of the force on the electrons.

circuit. If the attenuation is located several guide wavelengths from the tube's input after appreciable modulation of the beam has taken place, the forward-traveling wave is not attenuated nearly so much as the backward-traveling wave and substantially improved TWT performance results.

Traveling-wave tubes have been reported with noise figures of approximately 1 db, and noise figures under 6 db are not uncommon in the 2- to 4-Gc band. Noise figure tends to increase somewhat with frequency, and improved noise performance is expected at frequencies below 10 Gc. Although traveling-wave-tube noise has received a good deal of attention in the past, the subject is still imperfectly understood and an improved basic understanding may improve available noise figures in all frequency bands.

Traveling-wave tubes may be grouped in two general categories. In the first, the electrons are constrained to move parallel to the axis of the slow-wave circuit by some method of focusing, which often takes the form of a strong magnetic field parallel to the beam (and circuit) axis. Other forms of focusing sometimes used include periodic magnetic focusing and periodic electric focusing, in which a series of magnetic or electric lenses reconverge the electron beam, which tends to diverge because of the electronic space charge. In the second general category, the beam moves

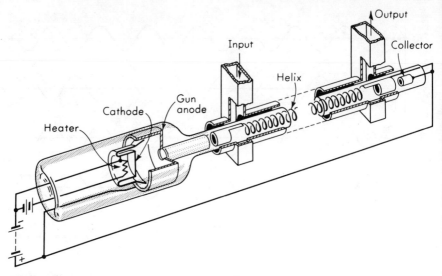

Fig. 3-26 Schematic of a traveling-wave tube. (*After Pierce.*)

through crossed electric and magnetic fields at a velocity given by the ratio of **E**, the electric field, to **B**, the magnetic flux density. The traveling-wave interaction is more complicated in this case, partly because the electronic equations of motion are more complicated. A typical example of the first and more common category of tube, generally called merely a traveling-wave tube, is shown in Fig. 3-26. A typical example of the second type, often called a crossed-field tube, or linear magnetron, is shown in Fig. 3-27.

One important difference between the fundamental energy relations

Fig. 3-27 Schematic diagram of a crossed-field traveling-wave tube.

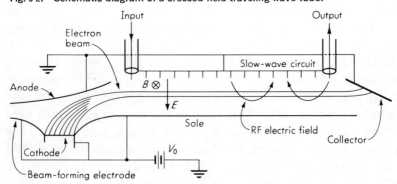

microwave communications

in these two types of traveling-wave tubes should be emphasized. The traveling-wave tube of Fig. 3-26 amplifies the electromagnetic wave at the expense of the *kinetic* energy of the electrons in the beam. If the tube is to have high efficiency, the electrons must be slowed down to a fraction of their initial velocity; unless special precautions are observed, the electron bunches will then lose synchronism with the circuit wave and the cumulative traveling-wave interaction will break down. There have been attempts to decrease the circuit velocity near the output end of this type of tube and to decrease the potential of the collector below the circuit potential to minimize the energy dissipated as heat, but even with these special techniques, the traveling-wave tube is not a fundamentally efficient device. Typical efficiencies of high-power traveling-wave tubes range from 20 to 40 percent.

The crossed-field tube, on the other hand, amplifies the circuit wave at the expense of the *potential* energy of the electrons. If an electron is slowed down, it moves closer to the anode (see Fig. 3-27), gains potential energy from the electric field which increases its velocity back to the synchronous velocity, and remains in step with the wave. Thus high-power crossed-field devices are capable of somewhat higher efficiencies, figures of from 40 to 70 percent being typical. The power output of these devices is limited by another consideration, however, especially at the higher frequencies. To achieve large power outputs, the electrons are allowed to approach very close to the circuit, and many of them strike it. As the frequency of operation increases, the dimensions of the circuit decrease and the circuit's ability to dissipate power also decreases. The electron-beam size must decrease, and to obtain the same power output, it follows that the power density in the beam must increase. Thus, just as the circuit is able to dissipate less power per unit area, the demands on it are increased. This same reasoning applies to the ordinary traveling-wave tube, except the beam does not inherently approach the circuit more closely as its energy is extracted. Thus by using stronger focusing to keep the beam from striking the circuit, more power can be put into the beam.

For many systems, weight is an important parameter which should be minimized. Early traveling-wave tubes were quite heavy when the weight of the solenoid which supplied the focusing magnetic field was included. Rather extensive development of lightweight packaging and focusing of traveling-wave tubes has led to dramatic improvements, and tubes which deliver watts of output power are now available in weights of less than 1 lb, including permanent magnets for focusing. This weight obviously does not include the power supply for running the tube.

Traveling-wave-tube Circuits

Traveling-wave-tube circuits that are generally used propagate a slow electromagnetic wave and are periodic. The helix has been the most popular circuit because of its ease of fabrication, high interaction impedance, and large bandwidth. Other circuits which have higher power-handling capabilities, or which are easier to scale to higher frequencies, have also become popular in recent years. Although any exhaustive discussion of the many circuits presently used in traveling-wave tubes is quite beyond the scope of this text, it is worthwhile to assess certain general characteristics of circuits used for traveling-wave tubes.

First, consider the helix. The phase velocity v of this circuit, normalized to the velocity of light, is shown in Fig. 3-28. At low frequencies, the circuit is quite dispersive; i.e., the phase velocity varies rapidly with frequency. At higher frequencies, the phase velocity is almost uniform over a wide band; in this frequency range, the wave can be considered as traveling down the wire with the velocity of light. For small pitch angles, the wave travels along the wire approximately one circumference to advance one pitch p of the helix; thus $v/c \approx p/2\pi a$. Another often-used diagram which shows the propagation characteristics of a periodic circuit particularly well is the $\omega\beta$ (or Brillouin) diagram shown in Fig. 3-29. In this diagram, the phase velocity at any point of the curve is merely the ordinate ω divided by the abscissa β. Furthermore, the group velocity of the wave is merely the slope of the curve $\partial\omega/\partial\beta$. This diagram has

Fig. 3-28 Helix normalized phase velocity versus frequency.

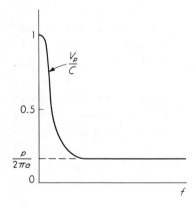

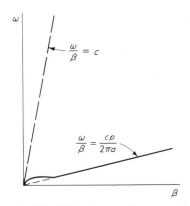

Fig. 3-29 Brillouin or $\omega\beta$ diagram for a helix.

even more interesting properties, as will now be demonstrated. Consider any periodic structure with characteristic period in the z direction of p. Consider also that the propagation characteristic β_0 is known for a given frequency range for this structure. The electric field which propagates on the structure may be written as

$$E = E_1(x,y,z)e^{j(\omega t - \beta_0 z)} \tag{63}$$

where β_0 is defined for the given frequency range and, in particular, is known for the radian frequency ω. Since the structure is periodic in the axial or $+z$ direction with period p, the complex vector amplitude of the electric field must also have the same periodicity p in the z direction and may be expanded in a Fourier series, as follows:

$$E_1(x,y,z) = \sum_{n=-\infty}^{\infty} E_{1n}(x,y) e^{-j\frac{2\pi n z}{p}} \tag{64}$$

Substituting this equation into the previous one, we may write the total electric field as

$$E = \sum_{n=-\infty}^{\infty} E_{1n}(x,y) e^{j\left[\omega t - \left(\beta_0 + \frac{2\pi n}{p}\right)z\right]} \tag{65}$$

The propagation constant of the nth term is seen from the last equation to be

$$\beta_n = \beta_0 + \frac{2\pi n}{p} \tag{66}$$

and the nth-order term of (65) is commonly called the nth-order *spatial harmonic* of the electric field in the periodic structure. The generality of the above result should be emphasized. Although the form of the electric field is still unknown, the propagation constants of all the spatial harmonics have been determined quite generally, assuming only that β_0 is known. If the electric field E is known, the complex vector amplitudes of any harmonic can also be determined by the usual formula

$$E_{1n}(x,y) = \frac{1}{p} \int_z^{z+p} E e^{-j(\omega t - \beta_n z)} \, dz \tag{67}$$

Next, consider the implication of (66) to the $\omega\beta$ diagram. If the propagation constant is known at any ω, the propagation constants of all spatial harmonics can be determined merely by knowing the periodicity p of the circuit. As an example, the $\omega\beta$ diagram for a helix, including several spatial harmonics, is shown in Fig. 3-30.

Note in particular the -1 spatial harmonic in Fig. 3-30. Because it occurs in the second quadrant of the $\omega\beta$ diagram, its phase velocity is negative. However, its slope is positive (and identical to the slopes of all other spatial harmonics at the same frequency), and its energy flow is therefore in the $+z$ direction. For traveling-wave interaction to occur, the electron beam must move with the *phase* velocity of the circuit wave. Suppose the energy on the circuit moves in the backward $(-z)$ direction, while the beam moves in the forward $(+z)$ direction. If this occurs and the beam velocity coincides with the -1 spatial harmonic's phase velocity, a special type of traveling-wave interaction is possible. In this case, the beam carries energy in one direction, the

Fig. 3-30 Brillouin diagram illustrating the spatial harmonics of a periodic circuit, the helix. Propagation is forbidden in the shaded region.

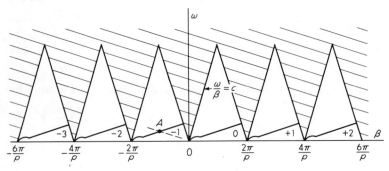

circuit carries energy in the opposite direction, and feedback occurs, leading to backward-wave oscillation. The frequency of this oscillation can be varied by changing the electron velocity and thus the intercept A in Fig. 3-30. Electronic tuning in this manner is possible over an octave range in some tubes; such electronic tuning, which is much more rapid than mechanical tuning, is extremely important in some microwave systems.

The shaded regions which occur at intervals of $2\pi/p$ along the β axis are regions in which propagation is forbidden for the following reason: The helix is an open structure, and a field analysis of any open periodic structure reveals that if the axial phase velocity of propagation of any spatial harmonic exceeds the velocity of light, the structure radiates energy. The nondispersive nature of the fundamental helix mode, taken with the presence of the forbidden regions of Fig. 3-30, implies that the helix will radiate if the circumference is equal to or greater than a free-space wavelength of the frequency in question. This observation, based on the properties of spatial harmonics, is verified by experiment. Other periodic structures which are used as traveling-wave-tube circuits are discussed by Beck,[12] who gives extensive references.

Small-signal Traveling-wave-tube Analysis

The purpose of the small-signal analysis of this section is to show that the signal propagating on the traveling-wave-tube circuit grows exponentially with distance in the presence of the electron beam, and also to explain the many simplifying assumptions necessary to derive this result. The impressive fact about this very simplified analysis is that it not only predicts exponential gain with distance along the circuit, but also does so with surprisingly good quantitative accuracy. It demonstrates that just because an exact analysis is difficult, one need not despair, but rather should set about performing the best approximate analysis possible under the circumstances. The approximate answer obtained should at the very least give insight into the problem, and perhaps will even give good quantitative results.

The problem to be solved may be stated as follows: Given a periodic circuit of finite length propagating a slow electromagnetic wave and an electron beam of finite diameter entering the circuit with velocity u_0 parallel to the axis of the circuit, determine the electric field, beam velocity, and beam current everywhere in the system, for a given power input at the start of the periodic circuit. Maxwell's equations are required, together with the force equation and the convection-current density equation. The force equation for a single electron is

$$\frac{d\mathbf{v}}{dt} = -\eta(\mathbf{E} + \mathbf{v} \times \mathbf{B}) \tag{68}$$

and the equation relating the convection-current density **J** to the charge density ρ and the electron velocity **v** is

$$\mathbf{J} = \rho \mathbf{v} \tag{69}$$

The above equations contain all the information necessary to solve for the exact fields which propagate along the circuit, once the boundary conditions of circuit and beam are known. In practice, these equations are too difficult to be solved exactly; instead, the following approximations are made:†

1. The electron beam is assumed to flow only in the axial direction; in practice, a large axial magnetic field restricts transverse motion of the beam, and this assumption is quite a good one. In equation form, this assumption becomes

$$\mathbf{v} = v\mathbf{a}_z \qquad \mathbf{J} = J\mathbf{a}_z \tag{70}$$

One consequence of this assumption follows from considering (68). Since the velocity v is in the $+z$ direction only, the $\mathbf{v} \times \mathbf{B}$ term in (68) cannot accelerate the electrons in the $+z$ direction and therefore this term can be neglected when the electron motion is determined.

2. Next, all quantities are assumed to be the sum of a time-invariant part and a small-amplitude time-varying part which has a wavelike variation in time and distance. Mathematically, this implies that

$$\begin{aligned} J &= -J_0 + J_1 e^{j\omega t - \Gamma z} \\ v &= u_0 + v_1 e^{j\omega t - \Gamma z} \\ \rho &= \rho_0 + \rho_1 e^{j\omega t - \Gamma z} \\ E_z &= E_1 e^{j\omega t - \Gamma z} \end{aligned} \tag{71}$$

where Γ is, in general, a complex number (for an unattenuated wave,

† This treatment follows the simplified treatment of J. R. Pierce, found in Ref. 13; his notation, which is generally accepted in the literature, has also been adopted.

Γ would be purely imaginary). The minus sign is introduced into the linearization of J in order that J_0 may be a positive number (the current flow in a TWT is in the $-z$ direction). The above linearization restricts our results to the small-signal case, but in this case, it reduces a nonlinear equation (69) to a dc equation plus a linear equation with wavelike variations in time and distance. Thus

$$-J_0 = \rho_0 u_0 \qquad J_1 = \rho_0 v_1 + \rho_1 u_0 \tag{72}$$

Here $e^{j\omega t - \Gamma z}$ is understood to multiply all terms of the second equation, and ρ_0 is a negative number since it is the electron charge density. The product $\rho_1 v_1$ has been neglected in (72) as a second-order term.

3. A time-varying electric field is generally expressed in terms of a scalar potential and the magnetic vector potential. In the case of slow-wave propagation, a straightforward exercise using Maxwell's equations shows that the electric-field contribution from the magnetic vector potential is $(v/c)^2$ times the contribution from the scalar potential. Since $v/c \ll 1$, the electric field can only be found from the scalar potential, which in turn can be related to an alternating voltage of the slow-wave circuit. Although the analysis of traveling-wave-tube interactions can be carried out in terms of electric field without introducing the circuit voltage, the latter procedure is followed below to emphasize the analogy between the traveling-wave-tube circuit and a transmission line, with which most readers are assumed familiar.

If an electric field with axial amplitude E_1 propagates on the circuit, it will alter the electron velocity in accordance with the force equation. Taking into account the above assumptions, the force equation (68) may be written as

$$\frac{dv}{dt} = \left(\frac{\partial}{\partial t} + \frac{dz}{dt}\frac{\partial}{\partial z}\right)v_1 = (j\omega - \Gamma u_0)v_1 = -\eta E_1 \tag{73}$$

where second-order terms have been neglected in accordance with the small-signal approximation. Likewise, the continuity equation can be written as

$$\nabla \cdot \mathbf{J} + \frac{\partial \rho}{\partial t} = -\Gamma J_1 + j\omega \rho_1 = 0 \tag{74}$$

Equations (72) to (74) involve four unknowns. If we choose to eliminate the charge density and velocity, the current density may be expressed in terms of the electric field as

$$J_1 = +j\frac{\omega}{u_0}\frac{\eta J_0}{(j\omega - u_0\Gamma)^2} E_1 \qquad (75)$$

If the electric field E_z has uniform amplitude over the cross-sectional area of the electron beam, the alternating current i will be proportional to the direct current I_0 with the same proportionality constant which exists between J_1 and J_0. The alternating current in the electron beam then is

$$i = j\frac{\beta_e I_0}{2V_0(j\beta_e - \Gamma)^2} E_1 \qquad (76)$$

where $\beta_e = \omega/u_0$ and corresponds to the phase shift per unit length of a disturbance which travels with the electron velocity. Equation (76) states that if an electric field E_z propagating on the slow-wave circuit modulates the electron beam, the alternating current on the beam can be obtained if all factors in the proportionality constant are known, including Γ. If Γ is known, the variation of all parameters with distance is determined and a given problem with specified boundary conditions can be solved.

So far the effect of the electron beam on the circuit has been neglected; by analyzing this effect, an additional equation between E_1 and i will be obtained. In any practical case, the current which induces an electric field on the circuit will be the current produced by the circuit's electric field in the first place. Therefore the equation to be derived and (76) will be two equations containing two unknowns, E_1 and i, and in addition, Γ. These equations can be solved for the propagation constant, since they determine it.

The equivalent circuit of Fig. 3-31 is commonly used for forward-wave traveling-wave tubes. The lossless transmission line is near an excited electron beam, which induces a current in the transmission line. The physical mechanism by which the current is induced can be seen by considering the closed cylindrical surface about the electron beam in Fig. 3-31. The current flowing into the left-hand end of the surface is i. The current flowing out of the right-hand end is $i + (\partial i/\partial z)\, dz$, where the length of the cylinder dz is assumed short compared with a wavelength of the

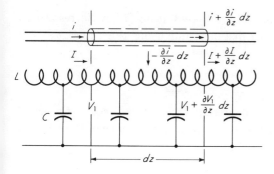

L = inductance per unit length
C = capacitance per unit length
i = alternating beam current
I = alternating current on transmission line
V_1 = alternating voltage on transmission line

Fig. 3-31 Schematic diagram of electron beam coupled to a slow-wave circuit represented by a transmission line.

wave along the transmission line. If $(\partial i/\partial z)\,dz$ is positive, more convection current comes out of the cylinder than went in and the alternating charge density inside the cylinder will be negative. Field lines from the transmission line to this negative charge density will constitute a displacement current. The value of this displacement current could be calculated, but this can be avoided by applying Kirchhoff's law to the total current (displacement plus convection) into the cylinder. If the total current into the cylinder is zero, then $-(\partial i/\partial z)\,dz$ flows out of the cylinder's sides and into the transmission line (this assumes no displacement current goes through the ends of the cylinder). The current from length dz of the inductance due to voltage V across the capacitance $C\,dz$ is $C\,dz\,(\partial V/\partial t)$, and therefore the Kirchhoff current law for the transmission-line section of length dz gives, after simplification,

$$\frac{\partial I}{\partial z} = -C\frac{\partial V}{\partial t} - \frac{\partial i}{\partial z} \tag{77}$$

Kirchhoff's voltage law around the closed loop gives V from ground to the inductance at z, $-L\,dz\,(\partial I/\partial t)$ from z to $z + dz$, and $-[V + (\partial V/\partial z)\,dz]$ from the inductance to ground at $z + dz$. Therefore, after simplification,

$$\frac{\partial V}{\partial z} = -L\frac{\partial I}{\partial t} \tag{78}$$

Eliminating the circuit current I by differentiating (77) with respect to time and (78) with respect to distance and using the exp $(j\omega t - \Gamma z)$ dependence yields

$$\Gamma^2 V = -\omega^2 LCV - j\omega L\Gamma i \tag{79}$$

If $i = 0$, the electron beam is not present and (79) reduces to the usual wave equation. The propagation constant in the absence of the beam is defined as Γ_0; from (79), with $i = 0$,

$$\Gamma_0 = j\omega \sqrt{LC} \tag{80}$$

The characteristic impedance of the line may be formed from (78) as

$$\frac{V}{I} = \sqrt{\frac{L}{C}} = Z_0 \tag{81}$$

When the beam is present, (79) may be rewritten as

$$i = -\frac{(\Gamma^2 - \Gamma_0^2)V}{\Gamma\Gamma_0 Z_0} \tag{82}$$

Equation (82) may be compared with (76), which has the beam alternating current written in terms of the electric field at the beam instead of the circuit voltage. Clearly the circuit voltage and the electric field are related; if the field is nearly independent of the transverse dimension of the circuit, for slow waves on the helix the axial electric field E_z is given by $-\partial V/\partial z = \Gamma V$. Using this auxiliary information, the ratio i/V, formed using (76) and (82), is

$$\frac{i}{V} = -\frac{(\Gamma^2 - \Gamma_0^2)}{\Gamma_0 \Gamma Z_0} = +j\frac{\beta_e \Gamma I_0}{2V_0 (j\beta_e - \Gamma)^2} \tag{83}$$

where $\beta_e = \omega/u_0$, the equivalent phase shift per unit length of an electron with velocity u_0, has been used. These two ratios have been equated in (83) because, for consistency, the beam current produced by the circuit voltage must be the same current which influences the circuit voltage. Equation (83) is a quartic equation in the propagation constant Γ, which it defines. This is seen best by writing it in the following form:

$$(\Gamma^2 - \Gamma_0^2)(j\beta_e - \Gamma)^2 = -j\frac{\Gamma^2 \Gamma_0 \beta_e I_0 Z_0}{2V_0} \tag{84}$$

If $I_0 \to 0$, then the roots are $\Gamma = \pm \Gamma_0$ and $\Gamma = j\beta_e$, that is, the normal propagation constants of the circuit and the propagation constant of the electron stream.

When I_0 is finite, the exact solution of (84) is best accomplished with numerical methods and a digital computer. However, an approximate solution is obtained relatively easily when the electron-beam velocity and the circuit phase velocity are identical. This is equivalent to setting $\Gamma_0 = j\beta_e$, and the resulting form is

$$(\Gamma - j\beta_e)^3(\Gamma + j\beta_e) = +2\Gamma^2\beta_e^2 \frac{I_0 Z_0}{4V_0} = +2C^3\Gamma^2\beta_e^2 \tag{85}$$

where C is the traveling-wave-tube gain parameter, as will be demonstrated below, and is defined as

$$C = \left(\frac{I_0 Z_0}{4V_0}\right)^{\frac{1}{3}} \tag{86}$$

For small values of C, the solution of (85) will be near the solutions obtained when C (or Z_0) is zero. Thus if Γ is written as

$$\Gamma = j\beta_e - \beta_e C \delta \tag{87}$$

where $C\delta \ll 1$, and this expression is substituted into (85), the following equation results:

$$(-\beta_e C \delta)^3(j2\beta_e - \beta_e C \delta) = +2\beta_e^2 C^3(-\beta_e^2 - 2j\beta_e^2 C\delta + \beta_e^2 C^2 \delta^2) \tag{88}$$

Since $C\delta \ll 1$, the quantity in the second set of parentheses on the left-hand side of the last equation is approximately $j2\beta_e$ and the quantity in parentheses on the right-hand side of the equation is approximately $-\beta_e^2$. The last equation then reduces to

$$\delta = -j^{\frac{1}{3}} = e^{-j(\pi/2 + 2n\pi)/3} \tag{89}$$

The three roots of this equation are labeled δ_1, δ_2, and δ_3; thus

$$\begin{aligned}
\delta_1 &= e^{-j\pi/6} = \frac{\sqrt{3}}{2} - \frac{j}{2} \\
\delta_2 &= e^{-j5\pi/6} = -\frac{\sqrt{3}}{2} - \frac{j}{2} \\
\delta_3 &= e^{-j3\pi/2} = j
\end{aligned} \tag{90}$$

By neglecting certain terms in (88), one root of the fourth-order polynomial has been lost; by letting $\Gamma = -j\beta_e - \beta_e C \delta_4$, this root, which corresponds to the backward-traveling circuit wave, is also found and is written as

$$\delta_4 = -j\frac{C^2}{4} \tag{91}$$

Thus the four values of Γ are

$$\begin{aligned}
\Gamma_1 &= j\left(\beta_e + \beta_e \frac{C}{2}\right) - \beta_e C \frac{\sqrt{3}}{2} \\
\Gamma_2 &= j\left(\beta_e + \beta_e \frac{C}{2}\right) + \beta_e C \frac{\sqrt{3}}{2} \\
\Gamma_3 &= j\beta_e(1 - C) \\
\Gamma_4 &= -j\beta_e\left(1 - \frac{C^3}{4}\right)
\end{aligned} \tag{92}$$

If a wave is identified with each value of Γ, the first two waves propagate more slowly than the circuit wave when the beam is absent and the first wave increases exponentially with distance while the second decays exponentially with distance. The third wave is unattenuated, but faster than the first two waves, while the fourth wave is the backward circuit wave, only slightly perturbed.

Now that the propagation constants in this special case have been determined, the boundary conditions on the beam at the input and output of the helix can be specified and the amplitude of the wave as a function of distance can be calculated; in particular, the amplitude of the wave at the output can be compared with the input amplitude, and the TWT gain can be derived. First, the output of the TWT is assumed matched to a load, and no power travels along the circuit in the $-z$ direction. (No power is reflected at the output end of the tube.) This is equivalent to stating that the wave corresponding to Γ_4 is not excited. The input voltage is assumed known and will be labeled V. Each of the three waves will have an associated voltage with distance variation given by $\exp(-\Gamma_i z)$, where $i = 1, 2,$ and 3. This is equivalent to stating that

$$\begin{aligned}
V(z) &= V_1 e^{-\Gamma_1 z} + V_2 e^{-\Gamma_2 z} + V_3 e^{-\Gamma_3 z} \\
&= \sum_{i=1}^{3} V_i e^{-\Gamma_i z}
\end{aligned} \tag{93}$$

The input current and input velocity provide the other boundary conditions. From (83) the current in each partial wave is given by

$$i_i = -\frac{I_0}{2V_0 C^2} \frac{V_i}{\delta_i^2} e^{-\Gamma_i z} \tag{94}$$

and the total current is

$$i = -\sum_{i=1}^{3} \frac{I_0}{2V_0 C^2} \frac{V_i}{\delta_i^2} e^{-\Gamma_i z} \tag{94a}$$

Similarly, from (73) the velocity can be shown to be

$$v_1 = -\sum_{i=1}^{3} j \frac{u_0}{CV_0} \frac{V_i}{\delta_i} e^{-\Gamma_i z} \tag{95}$$

Equations (93), (94a), and (95) provide the three necessary relations between V_1, V_2, and V_3, given the values of i and v_1 at the circuit input. (The reader hopefully is not confused by the notation. The subscripts 1, 2, and 3 on V, Γ, and δ refer to the roots of the propagation constant, and v_1 still refers to the fluctuating component of velocity of the total wave.)

In the usual case, v_1 and i are zero at the circuit input. The three equations for V_1, V_2, and V_3 are then given by (93), (94a), and (95), evaluated at $z = 0$. Thus

$$V = V_1 + V_2 + V_3$$
$$i = -\frac{I_0}{2V_0 C^2}\left(\frac{V_1}{\delta_1^2} + \frac{V_2}{\delta_2^2} + \frac{V_3}{\delta_3^2}\right) = 0 \tag{96}$$
$$v_1 = -j\frac{u_0}{2CV_0}\left(\frac{V_1}{\delta_1} + \frac{V_2}{\delta_2} + \frac{V_3}{\delta_3}\right) = 0$$

These may be solved for V_1, V_2, and V_3 by standard methods, giving

$$V_1 = V_2 = V_3 = \frac{V}{3} \tag{97}$$

and the total voltage along the circuit as a function of distance may be found by substituting this result into (93). When the circuit length L is sufficiently large so that $\beta_e C \delta_1 L$ is large compared with unity, the first wave predominates and

$$V(L) = \frac{V}{3} \exp\left(\frac{\beta_e C \sqrt{3}}{2} L\right) \exp\left[-j\left(\beta_e + \frac{\beta_e C}{2}\right)L\right] \tag{98}$$

Customarily, $\beta_e L$ is written as $2\pi N$, where N is the circuit length in electronic wavelengths, $\lambda_e = u_o/f$. The power output is proportional to $|V(L)|^2$, and the power input is similarly proportional to $|V(0)|^2$. The decibel power gain is thus

$$G = 10 \log \frac{|V(L)|^2}{|V(0)|^2} = 10 \log \left(\frac{\exp 2\pi \sqrt{3}\, CN}{9}\right) = A + BCN \tag{99}$$

where

$$A = 10 \log \tfrac{1}{9} = -9.54 \text{ db}$$

and

$$B = 10 \times 2\pi \sqrt{3} \log e = 47.3 \text{ db}$$

Equation (99) shows that for the case considered there is an initial loss at the circuit input of 9.54 db. This loss occurs because the input voltage splits into three waves of equal magnitude and the growing wave voltage is only one-third the total voltage at the input. The term BCN shows that the gain is proportional to length of the TWT (so long as the small-signal theory holds) and also proportional to the gain parameter C, which is a measure of the strength of interaction between beam and circuit.

The above calculation of TWT gain demonstrates the fundamentals of the procedure. A similar procedure may be used for the backward-wave oscillator[14] and for the severed helix.[13]

The noise figure of a TWT amplifier is calculated by knowing the amplitude of the velocity and current modulation on the beam due to statistical variations in the emission velocities of the individual electrons and the number of electrons emitted per unit time. These quantities are affected by the potential variation with distance in the electron gun, and by adjusting the gun potential, the values of v_1 and i at the input can be adjusted to make the TWT noise figure a minimum. Thus the method described above is generally useful for many different TWT calculations. Its basic limitation is the small-signal linearization, which invalidates it for high-power calculations, including saturation power output and efficiency. Numerical methods using digital computers must be employed for these calculations.

EXERCISES

3-23 Show that the contribution of the magnetic vector potential **A** to the electric field **E** is $(v/c)^2$ times that of the scalar potential ϕ.

Note: $\mathbf{B} = \nabla \times \mathbf{A}$ and $\mathbf{E} = -\nabla \phi - \dfrac{\partial \mathbf{A}}{\partial t}$.

3-24 Show that (97) follows from (96) and (90).

3-25 An electron beam is velocity-modulated by a cavity and then immediately afterward enters the circuit input of a traveling-wave tube which is matched at both ends and has *no* input voltage. Determine the output voltage of the TWT in terms of the amplitude of the velocity modulation at the input, v_1. (Assume the current modulation on the beam is zero at the input.)

3-26 For a given TWT circuit, how is the decibel gain related to the electron velocity? Should fast or slow beams be used for short, high-gain tubes? Explain the physical reasons behind your answer.

REFERENCES

Induced Current

1 SHOCKLEY, W.: Current Induced by a Moving Charge, *J. Appl. Phys.*, vol. 9, pp. 635–636, October, 1938.

2 RAMO, S.: Currents Induced by Electron Motion, *Proc. IRE*, vol. 27, pp. 584–585, September, 1939.

Transit-time Effects

3 LLEWELLYN, F. B., and L. C. PETERSON: *Proc. IRE*, vol. 32, pp. 144–146, March, 1944.

Klystrons

4 HEIL, A. A., and O. HEIL: A New Method of Generating Undamped Electromagnetic Waves of High Intensity, *Z. Physik*, vol. 95, p. 752, July, 1935.

5 HAHN, W. C., and G. F. METCALF: Velocity-modulated Tubes, *Proc. IRE*, vol. 27, pp. 106–116, February, 1939.

6 VARIAN, R. H., and S. F. VARIAN: A High Frequency Amplifier and Oscillator, *J. Appl. Phys.*, vol. 10, p. 321, May, 1939.

Beams

7 SPANGENBERG, KARL R.: "Vacuum Tubes," McGraw-Hill Book Company, New York, 1948.

8 PIERCE, JOHN R.: "Theory and Design of Electron Beams," 2d ed., D. Van Nostrand Company, Inc., Princeton, N.J., 1954.

Velocity Modulation

9 BECK, A. H. W.: "Velocity Modulated Thermionic Tubes," pp. 55–57, Cambridge University Press, Cambridge, 1948.

10 Ref. 9, pp. 42–43.

11 MCLACHLAN, N. W.: "Bessel Functions for Engineers," p. 43, eq. (11), Oxford University Press, London, 1934.

Traveling-wave Tube

12 BECK, A. H. W.: "Space-charge Waves," Pergamon Press, New York, 1958.

13 PIERCE, J. R.: "Traveling-wave Tubes," chap. 9, D. Van Nostrand Company, Inc., Princeton, N.J., 1950.

14 JOHNSON, H. R.: Backward Wave Oscillators, *Proc. IRE,* vol. 43, pp. 684–697, June, 1955.

4

principles of solid-state microwave devices

4-1. INTRODUCTION

Solid-state devices are becoming increasingly important for the generation and amplification of microwave signals. Two classes of solid-state devices will be discussed in this chapter, and an attempt will be made to examine their principles of operation. The first class includes semiconductor junction devices such as transistors, varactor diodes (used in parametric amplifiers), and tunnel diodes (which have a fundamental negative resistance). These devices are useful at microwave frequencies because the dimensions of their junctions are short compared to a wavelength (being of the order of 0.001 in. in a varactor diode down to about 100 Å for a tunnel diode) and the transit-time problem either is eliminated (as in the varactor diode) or is unimportant (tunneling is virtually an instantaneous occurrence).

The second class consists of those devices which operate by the stimulated emission of radiation and are generally termed quantum-electronic devices. This group includes the maser and the optical maser, or laser. It seems probable that amplifiers and oscillators based on the maser principle will be an extremely important source of coherent energy for communications systems in this region.

Other important solid-state devices that are not discussed in this chapter for lack of space include nonreciprocal elements, such as isolators, circulators, and gyrators, which are generally made with ferrites. Although these are not amplifying devices, they make possible certain types of amplifiers when used in conjunction with other devices, such as the varactor diode, and are important in systems applications.

A fourth class of solid-state amplifiers, which may become important as solid-state technology develops, utilizes bulk interactions between electric waves and electrons in solids. One example of this class of device is the acoustical-wave amplifier.

4-2. THE TUNNEL DIODE

The tunnel diode is a nonlinear semiconductor junction diode. The doping of both the p and n regions of the tunnel diode is very high, impurity concentrations of 10^{19} to 10^{20} atoms/cm^3 being used. These high impurity concentrations cause the energy levels of the conduction band on opposite sides of the junction to be separated by an energy greater than the energy gap which exists between the valence and conduction bands, as is shown in Fig. 4-1a. With zero applied bias, the Fermi level lies within the valence band of the p-type material and within the conduction band of the n-type material. The electrons which exist in the conduction band of the n-type material may in some cases possess sufficient energy to overcome the retarding field between the p and n regions, and diffuse to the p region as in an ordinary diode. Similarly, minority carriers in the p region (of which there are very few because of the large doping) may be swept by the junction field into the n region. However, there is another means of current flow between p and n regions when the valence band of the n region overlaps the conduction band of the p region in voltage; electrons can tunnel directly through the very narrow depletion layer at the junction (which typically is of the order of 100 Å thick). There are energy levels available for occupancy on both sides of the junction, and the tunneling currents are very high, many orders of magnitude above the diffusion current mentioned above, for example. The tunneling process can be explained in quantitative detail only by using quantum mechanics. Field emission is another example of tunneling which is finding important applications in electron devices.

The characteristic iv curve for a tunnel diode is shown in Fig. 4-1b;

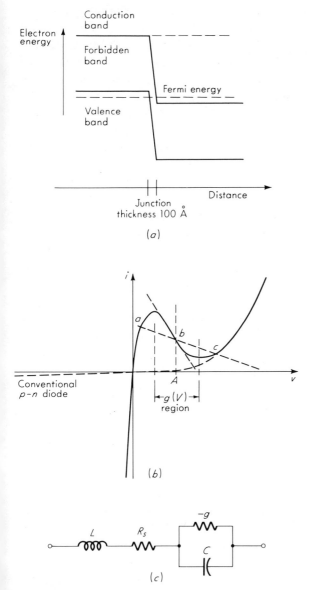

Fig. 4-1 Properties of the tunnel diode. (a) Band structure of tunnel diode at zero bias voltage; (b) characteristic curve of a tunnel diode; (c) equivalent circuit of a tunnel diode.

principles of solid-state microwave devices

the iv curve of a conventional p-n diode is shown dashed for purposes of comparison. Note that the reverse-bias current is very large and the forward-bias current has a negative slope (that is, a dynamic negative conductance) for values of forward bias in the range of a few tenths of a volt. If the left side of Fig. 4-1a is considered fixed (the reference voltage) and the right side considered variable, the forward bias implies raising the energy levels in the conduction band to the right of the junction and reverse bias implies lowering these energy levels. Consider the latter case. As the conduction band of the n material is lowered, still more energy levels become available for tunneling, and the net tunneling current increases rapidly as shown. However, as the conduction-band energy levels are raised, the overlap between the conduction band and the valence band decreases and finally ceases to exist. Similarly, the net tunneling current at first increases, and then decreases as the number of available energy levels decreases. Tunneling ceases as the voltage overlap between the bands goes to zero because the energy levels available for occupancy by a valence electron in the p material have also gone to zero. This is the reason the forward-bias part of the iv characteristic increases initially and then decreases. The second increase at higher voltages is due to the normal forward current which flows through any p-n junction diode.

Many important circuit applications of the tunnel diode can be discovered from its characteristic curve. Consider the abc load line; it intersects the characteristic curve in three points. Points a and c are stable points, and point b is unstable; i.e., if the current and voltage were given by point b and some small displacement from b occurred, the final value of i and v would be given by point a or c, but not by b. Since the tunnel diode has two stable states for this load line, it can be used as a binary memory device. Because of its very fast switching speeds, it is of great interest to computer engineers. However, it is also capable of microwave amplification or oscillation, and this property is of more interest here. Consider the second load line, which intersects the iv characteristic at point b only. This is a stable operating point and one at which the tunnel diode has a dynamic negative conductance, $-g = \dfrac{di}{dv}\bigg|_A$. For small voltage and current variations about the point A, the diode circuit behavior is adequately specified by $-g$. For larger signals, other approximations to the circuit conductance should be used. For many purposes, the above definition is sufficient, and the essence of the circuit behavior of the tunnel diode can be determined by using it in conjunction with the equivalent circuit shown in Fig. 4-1c.

The inductance is the lead inductance, which can be made as low as 10^{-9} to 10^{-10} henry with proper design. R_s is the "spreading" resistance due to bulk resistivity of the diode material; since the doping of tunnel diodes is so high, R_s is unusually low. C is the parasitic capacitance always found with p-n junctions and is of the order of 100 pf for a typical tunnel diode. The negative conductance has a value of $-g$ (g being the magnitude of the negative conductance). The ac equivalent circuit holds only for small signals at a bias point in the negative conductance region, where $-g = di/dv$ is evaluated at the operating point chosen (Fig. 4-1b).

The equivalent circuit in Fig. 4-1c can be replaced by a simpler circuit which facilitates analysis. Assuming that L is small enough to be neglected at the frequencies of interest, the admittance of the equivalent circuit is

$$Y(\omega) = \frac{-g(1 - R_s g) + (\omega C)^2 R_s + j\omega C}{(1 - R_s g)^2 + (\omega C R_s)^2} \tag{1}$$

which may be written as

$$Y(\omega) = \frac{-g(1 - R_s g)[1 - (\omega/\omega_c)^2] + j\omega C}{(1 - R_s g)^2 + (\omega C R_s)^2}$$

where

$$\omega_c = \frac{1}{C}\sqrt{\frac{g}{R_s}(1 - R_s g)}$$

Generally $R_s g \ll 1$, and under most conditions $(\omega C R_s)^2 \ll 1$, so the admittance reduces to

$$Y(\omega) = -g\left[1 - \left(\frac{\omega}{\omega_c}\right)^2\right] + j\omega C \tag{2}$$

The equivalent circuit for this approximate $Y(\omega)$ is a negative conductance shunted by a capacitance, as shown in Fig. 4-2. Over a narrow frequency range, the tunnel diode can be represented by a capacitance C

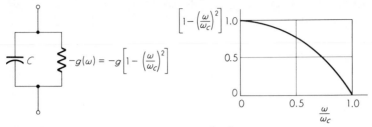

Fig. 4-2 Tunnel-diode modified equivalent circuit.

and a constant-shunt negative conductance whose value may be determined from Fig. 4-2.

A Tunnel-diode Amplifier

The circuit of Fig. 4-2 may be inserted between a source and load to provide electronic gain. The equivalent circuit of the resulting low-pass amplifier is shown in Fig. 4-3. Like any negative-conductance amplifier, a tunnel-diode amplifier is potentially unstable and will oscillate whenever the value of the load plus source conductance is less than the magnitude of the diode's negative conductance. If the isolated diode is considered as a two-port transducer, the power gain becomes infinite whenever oscillations occur, since the input power to the tunnel diode may be zero for finite output power. For this reason, the transducer power gain of the composite amplifier is generally used for tunnel diode and other negative-conductance devices. The load power P_L for the circuit of Fig. 4-3 is given as

$$P_L = \frac{1}{2} VV^* g_L = \frac{1}{2} \frac{II^* g_L}{(g_s + g_L - g)^2 + \omega^2 C^2} \tag{3}$$

and the transducer power gain is therefore

$$G_t = \frac{4 g_L g_s}{(g_s + g_L - g)^2 + \omega^2 C^2} \tag{4}$$

Fig. 4-3 Tunnel-diode amplifier equivalent circuit.

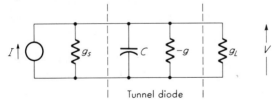

microwave communications

For low values of frequency and for values of $g_s + g_L - g$ very close to zero, the transducer gain can become very large. Note that it cannot go to infinity as long as the capacitor shunting the negative conductance has a finite value.

The low-frequency gain is a constant, and as the frequency is increased, the gain decreases. At frequencies sufficiently high, such that the capacitive term in the denominator of (4) predominates, G_t decreases by a factor of 4, or by 6 db, for each octave increase in frequency. The bandwidth is defined in this case as the frequency for which G_t is one-half its low-frequency value, and is given by

$$B = \frac{g_s + g_L - g}{2\pi C} \tag{5}$$

Note that as $g \to g_s + g_L$, the bandwidth $B \to 0$ even though the low-frequency transducer power gain approaches infinity. Thus gain can be traded for bandwidth and vice versa, and a good figure of merit for the amplifier is its gain-bandwidth product, which is usually defined as the maximum voltage gain of the amplifier times its bandwidth between 3-db points. The reader may have noticed that the voltage gain of this amplifier is unity, independent of the power gain; where power gain is the important quantity, it is customary to define the gain-bandwidth product as the square root of the transducer power gain (at zero frequency) times the bandwidth. Using this definition, the gain-bandwidth product for the tunnel-diode amplifier of Fig. 4-3 is

$$G_t^{\frac{1}{2}} B = \frac{\sqrt{g_s g_L}}{\pi C} \tag{6}$$

Note that the gain-bandwidth product is a constant, seemingly independent of the value of $-g$ but dependent upon the inherent capacitance associated with the diode junction. Increasing the source and load conductance would appear to increase the gain-bandwidth product, but to maintain the same value of transducer gain, the diode negative conductance would also have to be increased in magnitude.

The amplifiers of practical interest have appreciable gain, say, 10 db or more. In these cases, $g_s + g_L \approx g$, and the gain-bandwidth product is approximately $g_L(g - g_L)/\pi C$, which has a maximum value as a function of g_L at $g = 2g_L$. The maximum gain-bandwidth product is therefore approximately $g/2\pi C$ and under these conditions is a property of the tunnel diode alone. Values of $g/2\pi C$ in the tens-of-gigacycles range have been reported; these values will probably not be greatly improved, although precise circuit synthesis can be used to increase the bandwidth somewhat over the value given by (5).

The output power of a tunnel diode is limited, even though the transducer gain may become very large, by the definition of $-g$. In the equivalent circuit of Fig. 4-1c or Fig. 4-3, the value of $-g$ is taken as the slope of the characteristic curve (Fig. 4-1b) at the bias point (for example, point A). If the voltage variation about that point becomes large, the definition is no longer valid, and the magnitude of $-g$ decreases for large voltage swings. This limiting action reduces the gain for larger power inputs and is the physical limitation on the power output of a tunnel-diode oscillator. Because of the awkward, nonanalytic shape of the tunnel-diode characteristic curve, this limiting process is rather difficult to describe mathematically. However, it is similar in nature to the limiting that occurs in a reflex-klystron oscillator.

EXERCISE

4-1 A tuned tunnel-diode amplifier is obtained by adding a shunt inductance to the load in Fig. 4-3. Determine the transducer power gain as a function of frequency for this amplifier, and show that the gain-bandwidth product is the same as for the untuned amplifier.

4-3. PARAMETRIC AMPLIFIERS

Normal amplifiers convert power from a dc source (power supply or battery) into power at some signal frequency, i.e., time-varying or alternating power. On the other hand, a parametric amplifier converts power at one frequency (from a source generally called the pump) into power at another frequency, the signal frequency. Parametric amplifiers are of interest in microwave systems because they have a low noise figure and are relatively inexpensive to build and operate.

The pump voltage is mixed with the signal voltage by a nonlinear reactance, which in microwave systems is generally a varactor diode. The resulting equations describing the system are nonlinear and, as such, are difficult to solve. However, for small signals they can be reduced to linear equations with time-varying coefficients. This distinction is an important one since the solutions of linear equations with time-varying coefficients can be superimposed whereas the solutions of nonlinear equations cannot.

The physical principle involved in parametric amplification may be seen in the following example. Consider an LC circuit as shown in Fig. 4-4. In this circuit we shall assume that the capacitor plates can be physically pulled apart and pushed together. At $t = 0$, all the energy in the system is stored in the capacitance and the voltage across the capacitance is V_0. As time increases, the energy oscillates back and forth between the electric and magnetic fields of the system. When the voltage across the capacitor plates is zero, the plates are pushed together, which causes no change in the energy of the system. When the voltage across the capacitor plates is a maximum (either positive or negative), the plates are physically pulled apart. Since the presence of the electric field causes an attractive force to exist between the plates, pulling them apart does work on the system, i.e., builds up the energy in the system. In Fig. 4-4 the plates are pushed together at points on the time axis marked by a circle, and are pulled apart at points on the time axis marked by an ×. The energetics of this system is fairly easy to analyze. When the plates are suddenly moved, the charge Q on them does not change. If the relative magnitude of the capacitance change $|\delta C|/C \ll 1$, then the energy change

Fig. 4-4 Diagram illustrating how the electrical energy in a resonant LC circuit is increased by parametrically moving the capacitor plates.

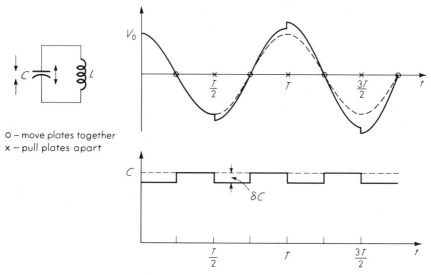

o – move plates together
× – pull plates apart

per half cycle of oscillation for the phase conditions shown in Fig. 4-4 is

$$\delta U_E \text{ per half cycle} = \frac{1}{2}\frac{Q^2}{C}\left(\frac{1}{1 - |\delta C|/C} - 1\right) = \frac{1}{2}\frac{Q^2}{C}\frac{|\delta C|}{C} \tag{7}$$

Since the energy stored in the capacitor is $Q^2/2C$, it follows that the change of energy per cycle in this system is

$$\delta U_E \text{ per cycle} = 2U_E \frac{|\delta C|}{C} \tag{8}$$

In any physical system some loss will be present, which may be represented by the quality factor Q_L of the system. Under steady-state conditions the energy lost per cycle must equal the energy input per cycle; the capacitance variation necessary to maintain steady-state conditions is therefore found from the definitions of quality factor. Thus

$$\begin{aligned}Q_L &= \frac{2\pi(\text{energy stored in circuit})}{\text{energy lost per cycle}} \\ &= \frac{2\pi U_E}{2U_E(|\delta C|/C)}\end{aligned} \tag{9}$$

Hence it follows that the relative capacitance variation needed for a system whose loss is described by the quality factor Q_L under abrupt changes in the energy storage element is given by

$$\frac{|\delta C|}{C} = \frac{\pi}{Q_L} \tag{10}$$

It is important to note in this example that the amplification is accomplished when a given phase exists between the pump frequency, which is twice the signal frequency, and the signal frequency. For example, if one pushed the capacitor plates together at the points on the time axis marked by ×'s and pulled them apart at points on the time axis marked by o's, the signal would be attenuated, not amplified.

EXERCISES

4-2 If the capacitance variation in the above example is sinusoidal in time, rather than abrupt, and $|\delta C|$ is the peak variation, determine the relative capacitance variation necessary to maintain the signal at an equilibrium level [this condition corresponds to (10) above].

4-3 Determine the variation of stored energy versus time for the abrupt capacitance-variation case, for initial energy U_{EO} and resonant frequency f.

The Manley-Rowe Relations

Two useful and general equations relating the power at different frequencies flowing into and out of a lossless nonlinear reactance have been derived by Manley and Rowe.[1] These general relationships are useful in determining whether or not power gain is possible in a reactive frequency conversion when the frequency of the signal is raised or lowered and in addition (since all physical nonlinear reactances have some loss), whether or not the power gain of a lossless nonlinear reactance is an upper bound of the gain which one can expect from a parametric amplifier using an actual reactance.

The model analyzed by Manley and Rowe is shown in Fig. 4-5. A voltage generator at frequency f_1 is in a series with its internal impedance, a bandpass filter centered at frequency f_1, and a nonlinear reactance. Similarly, a voltage source at frequency f_2 is in series with its internal impedance, a bandpass filter at f_2, and a nonlinear reactance. Separate loads are provided for each of the sum and difference frequencies generated by the mixing of frequencies f_1 and f_2 in the nonlinear reactance. Each of these load resistances is in series with a bandpass filter of the appropriate frequency. In Fig. 4-5 the sources are shown to the left of the nonlinear reactance, and the loads are shown to the right. An infinite number of loads should be shown if all sum and difference frequencies are considered. In Manley and Rowe's analysis, power flowing into the nonlinear reactance is considered positive, and power flowing out of the nonlinear reactance is considered negative.

Fig. 4-5 Circuit model used in deriving and understanding the Manley-Rowe relations.

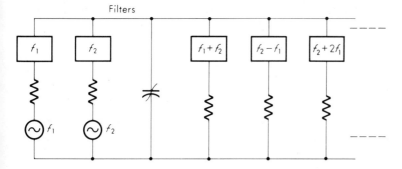

The equations this analysis obtains are as follows:

$$\sum_{m=0}^{\infty} \sum_{n=-\infty}^{\infty} \frac{mP_{m,n}}{mf_1 + nf_2} = 0 \qquad (11)$$

$$\sum_{m=-\infty}^{\infty} \sum_{n=0}^{\infty} \frac{nP_{m,n}}{mf_1 + nf_2} = 0 \qquad (12)$$

In these equations $P_{m,n}$ is the power at frequency $mf_1 + nf_2$ flowing into the nonlinear reactance. To demonstrate the utility and the usefulness of the Manley-Rowe relations, consider the case when all loads are zero except the one at frequency $f_1 + f_2$. Therefore power can be delivered from the nonlinear reactance to the load in series with the filter at frequency $f_1 + f_2$, and only these frequencies need be considered in the Manley-Rowe relations. Adopting the convention that frequency $f_3 = f_1 + f_2$ and that the power at frequency f_1 will be denoted by P_1, etc., the Manley-Rowe relationships in this case may be written as

$$\frac{P_1}{f_1} + \frac{P_3}{f_3} = 0 \qquad (13)$$

$$\frac{P_2}{f_2} + \frac{P_3}{f_3} = 0 \qquad (14)$$

Here P_1 is the power delivered from the source to the nonlinear reactance at frequency f_1, P_2 is the power delivered from the source to the nonlinear reactance at frequency f_2, and P_3 is the negative of the power absorbed by the load at frequency f_3. If f_1 is taken as the signal frequency and f_2 as the pump frequency, f_3 is the output frequency of this up-converter. The gain of the up-converter is the power delivered to the load at frequency f_3 divided by the power from the source at frequency f_1, and it is given by

$$G = -\frac{P_3}{P_1} = \frac{f_3}{f_1} \qquad (15)$$

Thus the up-converter power gain in this idealized case, using a lossless nonlinear reactance, depends only on the ratio of the output frequency and the input frequency.

EXERCISE

4-4 If f_2 is a pump frequency and f_3 (greater than f_2) is the input signal frequency, show that a down-converter with $f_1 = f_3 - f_2$ as the output frequency must necessarily have an output power less than the input power.

Methods of Analysis

The analysis of parametric devices involves either a nonlinear circuit element or a time-varying circuit element. Nonlinear analysis is extremely difficult, and approximations are generally utilized to make the analysis tractable. One approximation which is commonly used is the so-called small-signal approximation; this is the approach adopted here. The small-signal approach for parametric analysis assumes that the signal voltage appearing across the nonlinear element is much smaller than the pump voltage. In the case of low-noise parametric amplifiers, where the signal voltage is not much larger than noise signals and the pump voltage is considerably larger (sometimes as much as 100 db larger than the signal voltage), this assumption is almost always justified. We start by letting the signal voltage be v_1 and the pump voltage across the nonlinear element be v_2. Thus

$$v_1 = V_1 \cos \omega_1 t \qquad v_2 = V_2 \cos \omega_2 t \qquad (16)$$

where $V_1 \ll V_2$. Assuming that the nonlinear reactance is a capacitance, we write the charge on the capacitor plates in a Taylor series expansion about voltage v_2 and consider only the first two terms of the expansion. Thus

$$q(v_1 + v_2) = q(v_2) + \frac{\partial q}{\partial v}\bigg|_{v_2} v_1 \qquad (17)$$

In the above equation we can define the coefficient of v_1 as a capacitance,

$$C(v_2) \equiv \frac{\partial q}{\partial v}\bigg|_{v_2} \qquad (18)$$

Since v_2 is a periodic function, it follows that the capacitance defined above must also be a periodic function and therefore can be expressed as a Fourier series, as follows:

$$C(v_2) = \sum_{n=0}^{\infty} C_n \cos n\omega_2 t = C(t) \tag{19}$$

It is worth noting at this point that the small-signal approximation has reduced a nonlinear charge-voltage relationship expressed as a nonlinear capacitance to a time-varying capacitance.

The effect of a nonlinear reactance or a time-varying reactance with voltages impressed across it at two different frequencies is to produce harmonics of the impressed frequencies and also all the sum and difference frequencies, expressed as $mf_1 + nf_2$. Although an infinite number of new frequencies are produced in theory, in practice only a few frequencies are produced with appreciable amplitude, for in a practical parametric amplifier circuit, impedance sufficient to generate appreciable voltages is present only at a few discrete frequencies. In the analysis which follows it is assumed that the zero-order and first-order terms in the Fourier series for time-varying capacitance are the only terms with appreciable amplitude. Thus the capacitance variation may be written as

$$C(t) = C_0 + 2\gamma C_0 \cos \omega_2 t \tag{20}$$

Similarly, it is assumed that voltages may exist across a nonlinear capacitance only at the signal frequency and at the sum and difference frequencies which are given by $f_+ = f_2 + f_1$ and $f_- = f_2 - f_1$. Although currents can flow at other frequencies as well as those just described, the currents at other frequencies are short-circuited and can be ignored in our analysis. Thus the voltage across the nonlinear reactance and the currents which flow in the nonlinear reactance at the frequencies of interest may be expressed as

$$\begin{aligned} v &= V_1 e^{j\omega_1 t} + V_1^* e^{-j\omega_1 t} + V_+ e^{j\omega_+ t} + V_+^* e^{-j\omega_+ t} + V_- e^{j\omega_- t} + V_-^* e^{-j\omega_- t} \\ i &= I_1 e^{j\omega_1 t} + I_1^* e^{-j\omega_1 t} + I_+ e^{j\omega_+ t} + I_+^* e^{-j\omega_+ t} + I_- e^{j\omega_- t} + I_-^* e^{-j\omega_- t} \end{aligned} \tag{21}$$

Here, V_1, V_+, and V_- and the corresponding current amplitudes are phasor quantities, but v and i are real quantities, since they are composed of complex numbers plus their complex conjugates. To determine the impedance matrix which relates the phasor current quantities to the phasor voltage quantities, the fundamental equation relating current and charge is used, as follows:

$$i = \frac{d}{dt}[C(t)v(t)] \tag{22}$$

It should be particularly noted that both $C(t)$ and $v(t)$ in the last equation are real quantities. By substituting (20) and (21) into the last equation, the admittance matrix relating the current amplitudes to the voltage amplitudes is obtained. Thus

$$\begin{bmatrix} I_-^* \\ I_1 \\ I_+ \end{bmatrix} = \begin{bmatrix} -j\omega_- C_0 & -j\omega_- \gamma C_0 & 0 \\ j\omega_1 \gamma C_0 & j\omega_1 C_0 & j\omega_1 \gamma C_0 \\ 0 & j\omega_+ \gamma C_0 & j\omega_+ C_0 \end{bmatrix} \begin{bmatrix} V_-^* \\ V_1 \\ V_+ \end{bmatrix} \tag{23}$$

The last equation shows that the current at the signal frequency I_1 has components produced by the voltages at the sum and difference frequencies as well as the voltage at the signal frequency due to mixing in the nonlinear reactance. As a check on the method thus far, note that if γ is set equal to zero, the admittance matrix is diagonalized and the signal current is independent of the voltage at the sum and difference frequency, which is to be expected.

EXERCISE

4-5 Show that Eq. (23) follows from (20) to (22).

The Parametric Up-converter

The input signal in a parametric up-converter is at frequency f_1, and the output signal is at frequency f_+, where $f_+ = f_1 + f_2$; the frequency f_2, of course, is the fundamental frequency of the capacitance variation with time. The model used here in the parametric up-converter analysis is shown in Fig. 4-6. The filters at f_1 and f_+ pass currents only

Fig. 4-6 Circuit model used in deriving the gain of a parametric up-converter.

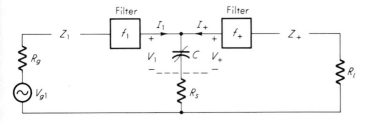

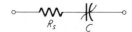

Fig. 4-7 Circuit model for a varactor diode.

at those frequencies. The signal source is a voltage generator with internal impedance R_g, and Z_1 represents the impedance at frequency f_1 of the microwave circuitry. The load resistance is R_L, and Z_+ represents the impedance at frequency f_+ of the microwave circuitry. The reactor diode is represented by a nonlinear capacitance in series with the spreading resistance R_s (Fig. 4-7). We assume as before that currents at frequencies other than f_+ and f_1 are shorted so that voltages produced by nonlinear mixing in the capacitance at these other frequencies are not generated across the capacitance. Since the circuit of Fig. 4-6 is a series circuit for the current at f_1 and also the current at f_+, it is more useful to use an impedance matrix instead of the admittance matrix derived above. The impedance matrix for the parametric up-converter may be obtained by setting V_-^* equal to zero in (23) and inverting the resulting matrix relating I_1 and I_+ to V_1 and V_+. When this is done, the following equation between voltage and current results:

$$\begin{bmatrix} V_1 \\ V_+ \end{bmatrix} = \begin{bmatrix} \frac{1}{j\omega_1 C} & \frac{-\gamma}{j\omega_+ C} \\ \frac{-\gamma}{j\omega_1 C} & \frac{1}{j\omega_+ C} \end{bmatrix} \begin{bmatrix} I_1 \\ I_+ \end{bmatrix} \qquad (24)$$

where $\omega_+ = \omega_1 + \omega_2$ and $C = C_0(1 - \gamma^2)$. The voltages at V_1 and V_+ are those which appear across the nonlinear reactance of the parametric diode. The circuit model may now be modified to that shown in Fig. 4-8, where the impedance elements inside the box refer to those in (24). This modified circuit model leads to the following relationship between

Fig. 4-8 Modified circuit model of up-converter; only currents at frequency f_1 flow to the left of the nonlinear capacitance (represented by the impedance matrix), and only currents at frequency f_+ flow to the right (through the load).

microwave communications

the generator voltage at frequency f_1 and a hypothetical generator voltage at frequency f_+, which has been inserted in Fig. 4-8 for the sake of symmetry. Thus

$$\begin{bmatrix} V_{g1} \\ V_{g+} \end{bmatrix} = \begin{bmatrix} Z_{T1} + Z_{11} & Z_{12} \\ Z_{21} & Z_{T+} + Z_{22} \end{bmatrix} \begin{bmatrix} I_1 \\ I_+ \end{bmatrix} \qquad (25)$$

In this last equation, $Z_{T1} = R_s + R_g + Z_1$, and $Z_{T+} = R_s + R_1 + Z_+$.

The transducer power gain for the parametric up-converter is a ratio of the power delivered to the load at frequency f_+ divided by the power available from the source at frequency f_1. This may be written as

$$G_t = \frac{P_L}{P_{\text{av1}}} = \frac{4R_L R_G I_+ I_+^*}{V_{g1} V_{g1}^*} \qquad (26)$$

To eliminate the currents and voltages from the last equation, we need a relationship between I_+ and V_{g1}, which may be obtained from (25) by setting $V_{g+} = 0$ and solving the equations (by eliminating I_1). The result of this operation is

$$I_+ = \frac{-Z_{21} V_{g1}}{(Z_{T1} + Z_{11})(Z_{T+} + Z_{22}) - Z_{12} Z_{21}} \qquad (27)$$

In order to maximize the current I_+ flowing to the load, in practice the impedances Z_1 and Z_+ are tuned to cancel the reactance of the parametric diode. This is equivalent to setting the imaginary part of $Z_{T1} + Z_{11} = 0$ and the imaginary part of $Z_{T+} + Z_{22} = 0$. When this is done, the transducer power gain is expressed by

$$G_t = \frac{4R_g R_L \gamma^2}{[\omega_1 C R_{T1} R_{T+} + (\gamma^2/\omega_+ C)]^2} \qquad (28)$$

The above expression for transducer power gain is not in a very useful form, since too many independent resistances appear. If the circuit losses are small compared with the diode losses, R_{T1} may be set equal to $R_g + R_s$ and R_{T2} may be set equal to $R_L + R_s$. Under these assumptions the transducer power gain may be written as

$$G_t = \frac{4R_g R_L \gamma^2}{[\omega_1 C (R_g + R_s)(R_L + R_s) + (\gamma^2/\omega_+ C)]^2} \qquad (29)$$

This last expression is more useful, since it is symmetrical in R_g and R_L. Therefore the same value of load resistance R_L, which maximizes the transducer power gain, will also be the value of R_g which maximizes the transducer power gain. If R_g is set equal to R_L and the resulting expression for transducer power gain is then maximized using well-known techniques of differential calculus, the following value for R_g results:

$$R_L = R_g = R_s \left(1 + \frac{\gamma^2}{\omega_1 \omega_+ C^2 R_s^2}\right)^{\frac{1}{2}} = R_s \left(1 + \frac{\omega_1}{\omega_+} \gamma^2 Q^2\right)^{\frac{1}{2}} \quad (30)$$

where the Q of the parametric diode, defined as $Q = 1/\omega_1 C R_s$, has been used in the last term. If this value for $R_L = R_g$ is substituted back into the expression for transducer power gain, the following results:

$$G_t = \frac{\omega_+}{\omega_1} \frac{x}{(1 + \sqrt{1 + x})^2} \quad (31)$$

where

$$x = \frac{\omega_1}{\omega_+} \gamma^2 Q^2$$

It was shown, using the Manley-Rowe relationships, that the maximum power gain one can expect with a lossless nonlinear reactance in a parametric up-converter is ω_+/ω_1. Therefore the factor multiplying ω_+/ω_1 in (31) is a degradation factor due to the finite loss of the parametric diode (Fig. 4-9). Note that as x goes to infinity, which implies that the loss in the diode decreases toward zero, the degradation factor goes to unity; in this limit one can achieve the gain predicted by the Manley-Rowe relations. Note also that if ω_+ becomes very large com-

Fig. 4-9 Degradation factor $x/(1 + \sqrt{1 + x})^2$ plotted versus x.

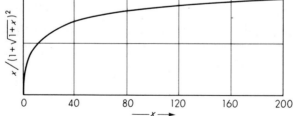

microwave communications

pared with ω_1, the transducer power gain tends toward $(\gamma Q)^2/4$. From this result and the fact that x is a function only of the frequency, the diode parameters, and Q, we surmise that a good figure of merit for the semiconductor parametric diode is γQ.

EXERCISES

4-6 Verify that when the loads and generator resistances are equal, their optimum value is given by (30).

4-7 Show that the transducer power gain is given by (31) when the optimum value of load resistance is used.

4-8 Suppose the input frequency is ω_+ and the output frequency is ω_1. Determine the transducer power gain in this case, and compare it with that which is expected from the Manley-Rowe relations.

4-9 Carry through an analysis for a difference-frequency parametric amplifier, where $f_2 - f_1 = f_-$. Can power gain be achieved at frequency f_1 when the input signal is at f_1? Use the circuit given in Fig. 4-10.

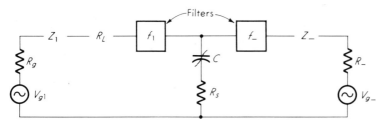

Fig. 4-10 Circuit diagram for a microwave parametric amplifier.

4-4. MASERS

Microwave amplification by the stimulated emission of radiation is a quantum-mechanical process performed by masers. These amplifiers add almost no noise to the signal they amplify, and hence they are an important preamplifier for microwave systems. A maser was used as the preamplifier for the radar signal returned from Venus (the "Venus bounce"), and masers make transatlantic and transpacific satellite communication not only possible, but commercially desirable. There are several types of masers, including the ammonia maser, often used as a frequency standard; two-level masers, suitable only for pulsed operation; and three- (or more) level masers, suitable as continuous amplifiers (CW amplifiers). The three-level type of device is often much more desirable for a communication system than a pulsed amplifier. We shall discuss the paramagnetic three-level maser as a means of understanding microwave amplification by the stimulated emission of radiation.

The reader is assumed to understand the basic concept of modern atomic physics, which holds that electrons bound to an atom must occupy discrete energy levels. For an electron to change from one energy level to another, the atom must absorb (or radiate) a photon of energy,

$$E = hf \tag{32}$$

where h is Planck's constant

$$h = 6.625 \times 10^{-34} \text{ joule-sec} = 4.135 \times 10^{-15} \text{ ev-sec}$$

and f is the frequency of the incident photon. Energy levels of atoms are generally spaced by an electron volt or so, which (as Fig. 4-11 shows) corresponds to photons of visible light.

The material most often used for three-level masers is ruby, which is Al_2O_3 with about 0.1 percent chromium added. The chromium atoms occupy aluminum atom sites in the ruby crystal lattice, but their magnetic

Fig. 4-11 Energy level separation versus radiated frequency.

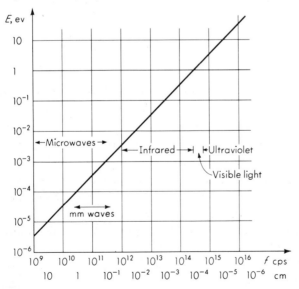

moments or spins are randomly oriented in the absence of an external magnetic field. When an external magnetic field is applied, the spins tend to align with this field. The net spin of the Cr^{3+} ion is due to unpaired electrons in the d shell of the atom, each of which has a spin of $\frac{1}{2}$. If all are aligned with the applied field, the ion spin is $\frac{3}{2}$; if all are aligned against the field, the ion spin is $-\frac{3}{2}$. Other possible values of the ion spin are $\pm\frac{1}{2}$. Only these spins are allowed, and with each spin is associated an energy level for the Cr^{3+} ion. The energy levels for ruby as a function of the magnetic field are shown in Fig. 4-12 for a specific orientation. Two points are important here. The energy levels are spaced by a few gigacycles to a few tens of gigacycles, and the spacing can be varied by altering the magnetic field. Since the frequency of operation will correspond to separation between energy levels, the second point implies that magnetic tuning is possible.

We now postulate a model for the maser material (ruby, in this case). It is an assemblage of a large number of spins, approximately 10^{21} for a typical crystal. These spins are independent of one another. In the presence of an external magnetic field, any one spin may exist in one of four discrete energy levels. When all the spins are considered, the populations of the energy levels at thermal equilibrium may be specified by Maxwell-Boltzmann statistics. If the energy levels are numbered as in Fig. 4-12, N_1 is the population of energy level 1 and E_1 is the energy of

Fig. 4-12 Zeeman energy levels of the Cr^{3+} ion in ruby for a magnetic field applied at 90° to the symmetry axis of the ruby crystal. (*After Siegman.*)

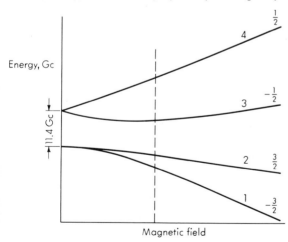

that level (similarly for levels 2, 3, and 4). Then at thermal equilibrium

$$\frac{N_1}{N_2} = e^{-(E_1-E_2)/kT} = e^{hf_{12}/kT}$$

$$\frac{N_2}{N_3} = e^{-(E_2-E_3)/kT} = e^{hf_{23}/kT} \tag{33}$$

$$\frac{N_1}{N_3} = e^{-(E_1-E_3)/kT} = e^{hf_{13}/kT}$$

A plot of energy level versus population N is shown in Fig. 4-13a for thermal equilibrium. A spin may be excited from a lower level to a

Fig. 4-13 (a) Population of energy levels at thermal equilibrium; (b) population of energy level with ruby pumped to saturation at frequency f_{13}, illustrating inversion between levels 2 and 1.

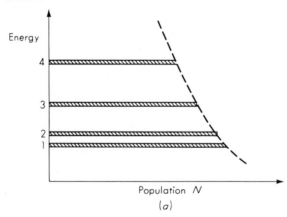

(a)

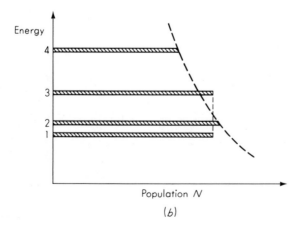

(b)

higher level thermally, i.e., by thermal lattice vibrations, etc., and may similarly change from a higher to a lower level.

There are two types of radiative transitions, induced transitions and spontaneous transitions. An induced transition occurs when the incident photon has the correct phase to excite an electron transition to a higher energy level (the photon is absorbed) or to excite an electron transition to a lower energy level (a photon is radiated, and whereas one photon was incident upon the atom, two are found leaving it *with the same phase*). A spontaneous transition occurs when an electron jumps to a lower energy level and emits a photon with no inducement.

At microwave frequencies, spontaneous transitions may be neglected compared with stimulated transitions, for any reasonable maser output power. The probability of stimulated transitions per unit time from level 1 to level 2 is here defined as W_{12} and is proportional to the signal power at frequency $f_{12} = (E_2 - E_1)/h$. The probability of transitions per unit time from level 2 to level 1 is W_{21}, and a quantum-mechanical analysis shows $W_{12} = W_{21}$. We may now write the power radiated by the spins as

$$P_{\text{rad}} = hf_{12}(W_{21}N_2 - W_{12}N_1) = hf_{12}W_{12}(N_2 - N_1) \tag{34}$$

where N_2 is the number of spins at energy E_2 and N_1 is the number at energy E_1. These numbers are related by Maxwell-Boltzmann statistics at thermal equilibrium as follows:

$$N_2 = N_1 e^{-(E_2-E_1)/kT} \tag{35}$$

Thus at thermal equilibrium power is always absorbed, since $N_2 < N_1$. A normal population of states is shown in Fig. 4-13a, corresponding to (33). Bloembergen suggested using three energy levels and, by pumping with a power source at frequency $f_{13} = (E_3 - E_1)/h$, for example, causing an inversion level either between state 2 and state 1 or between state 3 and state 2. The former inversion level is shown in Fig. 4-13b. The inversion is accomplished by radiating the maser material with many photons at energy hf_{13}. Since the population of level 1 is greater than that of level 3 and the transition probabilities are equal, more spins are pumped to level 3 than are induced to jump to level 1. Therefore, the populations of levels 1 and 3 rapidly become equal, corresponding to saturation.

With the inverted population of Fig. 4-13b, $N_2 > N_1$, and a weak signal at frequency f_{12} will induce more downward transitions than upward and will therefore be amplified. The signal power is assumed much smaller in amplitude than the pump power (at f_{13}); therefore, $W_{12} \ll W_{13}$, since these induced transition probabilities are proportional to the powers at their respective frequencies f_{12} and f_{13}. This allows us to assume that the population of level 2 is independent of the signal power but is dependent only on the saturation of the maser material at frequency f_{31} and the thermal relaxation process which allows spins to change energy levels because of thermal interactions with the crystal lattice.

It is apparent that the above treatment is very approximate. It does, however, give the essence of maser operation. A pump signal is necessary to saturate two energy levels so their population is equal. An inverted population can then exist in the crystal, leading to a net power gain due to induced transitions between the inverted levels.

One significant point should be stressed. In order that N_1/N_2 may be different from unity at thermal equilibrium, hf_{12}/kT should be of the order of unity. This means that at 10 Gc, T must be of the order of 1°K, for example. Thus very low temperatures are necessary for maser operation; without low temperatures, inversion would be impossible at microwave frequencies. This has advantages and disadvantages. On the plus side, at a few degrees Kelvin there is very little thermal noise, and thus the primary source of noise in a maser amplifier is the spontaneous radiative transitions, which have a very low probability. Therefore, the maser is a very low-noise amplifier. On the minus side, a maser generally needs liquid helium for a coolant. A system using a maser must therefore provide for a continuous supply of liquid helium. This requirement is not too difficult for an earth-based receiver. To put a maser into a satellite, however, would require a cryostat (to manufacture the liquid helium). Although this is not impossible, it is an important practical consideration when the maser is considered in a system's application.

The Laser

Light amplification by the stimulated emission of radiation is accomplished by the laser, a quantum-mechanical amplifier at visible light frequencies analogous to the maser. Different energy levels with a much greater separation must be used for the laser, of course (see Fig. 4-11), and relations first derived by Einstein show that spontaneous emission can no longer be neglected when compared with induced emission. Since the frequency of a photon of visible light is about 10^5 times the frequency of a microwave photon and since the power output of a quantum-mechanical amplifier is proportional to the number of induced emissions per second, one quickly concludes that for the same number of induced emissions per second the power output of the laser will be about 10^5 times as high as that of the maser. Furthermore, of the order of 10^5 times more power will be required to maintain a saturated condition between two energy levels in a laser than in a maser. This factor of 10^5 indicates that the thermal dissipation in the device will be increased by about the same factor and there may be a problem in keeping the device cool (and this is actually the case).

The perceptive reader may be wondering what produces the light signal which the laser amplifies. A light bulb, for example, emits visible light which is *incoherent;* i.e., the phases of the emitted photons are random (a photon has an amplitude, related to its frequency, and a phase). In addition, the light emitted by a light bulb has a broad spectral distribution (which is why it is white). White light, then, is completely analogous to noise, which will be studied in more detail in a later chapter. How, then, is a monochromatic (single frequency) coherent (all photons in phase) beam of light produced? Such a beam of light is necessary if we are to transmit information on a carrier wave at light frequencies, which is the purpose of communication. The generation of coherent visible light became possible only with the invention of the laser; like any other amplifier, a fraction of the output power can be fed back, leading to an unstable amplifier, i.e., an oscillator, which generates a coherent signal.

A typical laser oscillator is shown in Fig. 4-14. A 1-cm-diameter ruby rod 1 cm long is coated on one face with a totally reflecting layer of silver and on the other with a partially reflecting layer of silver. The other four faces are left transparent. The ruby is illuminated from the side by a very intense light pulse (usually from a xenon flash tube). This incoherent radiation from the flash tube serves as the pump signal and excites electrons from the ground state of ruby into one of two optical bands. Electrons can then give thermal energy to the lattice and drop into the so-called R levels. Their lifetime in these levels is relatively long, and if the crystal is pumped to saturation, the R levels are inverted with respect to the ground state so that a light signal may induce radiative transitions from the R levels to the ground state and thus be ampli-

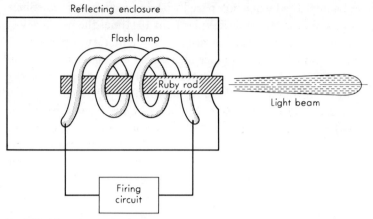

Fig. 4-14 The pulsed ruby rod laser.

fied. In the oscillator described above, the triggering mechanism is the spontaneous emission of photons from the R levels to the ground state. These spontaneously emitted photons may radiate out in any direction from the point of emission, all directions being equally probable. Some photons will strike the silvered or partially silvered end at right angles and be reflected back through the crystal again and again. These ends of the crystal are planes which must be very accurately parallel to each other. As the photons bounce back and forth along the same path, they induce other coherent emissions and thus the stable oscillation builds up. Part of the light is transmitted by the partially silvered end of the crystal, and this is the coherent light beam which is suitable for communications experiments. For example, as we shall see in the next chapter, if light of 5,000 Å wavelength were used in the above experimental setup, we would expect over half of the radiated energy to be concentrated in a half angle of about 5×10^{-5} rad, on the basis of very simple reasoning. This would produce a spot only 1 yd in diameter at a distance of 10 miles. If sensitive light amplifiers and detectors were available, a very high-resolution radar would be possible; one of the biggest problems would be how to aim the radar to hit a given spot (the problem is approximately the same as hitting the head of a nail with a bullet from a high-powered rifle from a distance of 100 yd). Such directivity also is appealing for nonjammable, private communication channels in outer space. (Such directivity has not been achieved yet experimentally, but half angles of about 10^{-2} rad have been reported by Maiman at Hughes Aircraft Company.)

It is important to emphasize that communication at light frequencies is only in an embryonic state. The first breakthrough, as outlined above, was the coherent light oscillator. All the other components necessary for a sophisticated communication system are still to be developed. Proposals have been made for modulators, detectors, etc., and work has commenced on these components. It is probable that waveguides as such will be used very little at light frequencies, since a 1-cm-diameter antenna may have the directivity mentioned above. Above all, the basic limitations of light amplifiers must also be established. However, the possibilities of communication at light frequencies are great. Consider the information capacity of a light amplifier with a bandwidth of 10^{-5} times its center frequency (a Q of 10^5). Let the center frequency be 2×10^{15} cps. The light amplifier will then pass a band of frequencies which is as wide as the band from 0 cps to 20 Gc, nearly the total bandwidth used for most communication purposes today!

REFERENCES

1   MANLEY, J. M., and H. E. ROWE: Some General Properties of Nonlinear Elements—Part I, General Energy Relations, *Proc. IRE,* vol. 44, pp. 904–913, July, 1956.
2   HEFFNER, H.: Solid-state Microwave Amplifiers, *IRE Trans. on Microwave Theory and Techniques,* vol. MTT-7, pp. 83–91, January, 1959. Contains an extensive bibliography.
3   HEFFNER, H.: Masers and Parametric Amplifiers, *Microwave J.,* vol. 2, pp. 33–40, March, 1959.
4   MOUNT, E., and B. BEGG: Parametric Devices and Masers: An Annotated Bibliography, *IRE Trans. on Microwave Theory and Techniques,* vol. MTT-8, pp. 222–243, March, 1960.
5   REED, E. D.: Variable-capacitance Parametric Amplifier, *IRE Trans. on Electron Devices,* vol. ED-6, pp. 216–224, April, 1959.
6   HEFFNER, H., and G. WADE: Minimum Noise Figure of a Parametric Amplifier, *J. Appl. Phys.,* vol. 29, p. 1262, August, 1958.
7   UHLIR, A.: The Potential of Semiconductor Diodes in High Frequency Communications, *Proc. IRE,* vol. 46, pp. 1099–1116, June, 1958.
8   TIEN, P. K., and H. HEFFNER: Parametric Amplifiers, *J. Res. Natl. Bur. Std.,* vol. 64D, pp. 751–754, November–December, 1960.
9   VINCENT, B. T.: A High-performance X-band Parametric Amplifier, *Proc. IRE,* vol. 49, pp. 511–512, February, 1961.
10   UENOHARA, M., and A. E. BAKANOWSKI: Low-noise Parametric Amplifier Using Germanium p-n Junction Diode at 6 kMc, *Proc. IRE,* vol. 47, pp. 2113–2114, December, 1959.

11 UENOHARA, M., and W. M. SHARPLESS: An Extremely-low-noise 6-kMc Parametric Amplifier Using Gallium Arsenide Point-contact Diodes, *Proc. IRE*, vol. 47, pp. 2114–2115, December, 1959.

12 SINGER, J. R.: "Masers," John Wiley & Sons, Inc., New York, 1959.

13 MEYER, J. W.: Systems Applications of Solid-state Masers, *Electronics*, vol. 33, no. 45, pp. 58–63, November 4, 1960.

14 BUDDENHAGEN, D. A., B. A. LENGYEL, F. J. MCCLUNG, JR., and G. F. SMITH: An Experimental Laser Radar, *IRE Intern. Conv. Record*, vol. 9, pt. 5, pp. 285–290, 1961.

5

antennas

5-1. INTRODUCTION

One of the purposes of a communication system is to propagate energy with which to transmit information between a source and the destination. Across short distances, this information may be transmitted directly by acoustic waves or by means of electric circuits such as transmission lines (two-wire lines, coaxial lines, waveguides, etc.). On the other hand, very much communication is carried out using wave propagation through space. The propagation of electromagnetic energy through space will be discussed in the next chapter. The device which couples the transmitter or receiver network to space we shall call the antenna.

To be really efficient, an antenna must have dimensions that are comparable with the wavelength of the radiation in which we are interested (see Fig. 5-1). At long wavelengths such as the part of the spectrum used in broadcasting (a frequency of 1 Mc/sec corresponds to a free-space wavelength λ of 300 m) the requirement on size poses severe structural problems, and it is consequently necessary to use structures that are fractions of a wavelength ($\lambda/8$ or $\lambda/4$). Such antennas have been aptly described as being little more than quasi-electrostatic probes protruding from the earth's surface. In order to control the spread of the energy, it is possible to combine antennas into arrays. We shall

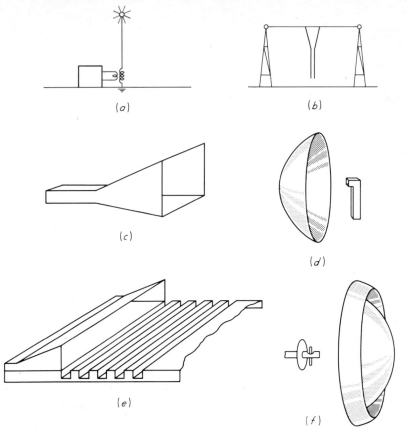

Fig. 5-1 Various types of antennas. (a) Top-loaded vertical mast; (b) center-fed horizontal antenna; (c) horn radiator; (d) paraboloidal reflector with a horn feed; (e) corrugated-surface wave system for end-fire radiation; (f) zoned dielectric lens with a dipole-disk feed.

treat some of the elements of array theory in a later section. As the wavelength gets shorter, it is possible to increase the size of the antenna relative to the wavelength; proportionately larger arrays are also possible, and techniques that are familiar in acoustics and optics can be employed. For example, horns can be constructed with apertures that are large compared with the wavelength. The horn can be designed to make a gradual transition from the transmission line, usually in this case a single-conductor waveguide, to free space. This results in broadband impedance characteristics as well as directivity in the distribution of energy in space. Another technique is to use an elemental antenna such as a horn or dipole together with a reflector or lens. The elemental antenna is essentially a point source, and the elementary design problem is the optical one of taking the rays from a point source and converting them into a beam of parallel rays. Thus one constructs a radio searchlight by using a paraboloidal reflector or a lens. A really large-scale structure of this basic form used as a receiving antenna (together with suitably designed receivers) serves as a radio telescope.

5-2. TYPES OF ANTENNA PROBLEMS

The antenna problem may arise in the following forms:

1. Given a specific radiating system, it is required that we find its directional characteristics, polarization of fields, gain, input impedance, and other important properties which need to be known in order that the communication system may be designed. In general, this problem can be solved using Maxwell's equations and the appropriate boundary conditions. In almost all cases, this is not an easy task. However, if a suitable current distribution on the conducting surfaces (as in the case of a thin half-wave dipole) or a suitable field distribution over an aperture (as in the case of a horn) can be assumed, the approximate solution may become simple to obtain and the results quite useful.

2. Given a directional characteristic, e.g., a certain beamwidth, polarization, and/or beam orientation, it is required that we find or select a radiating system which will produce it. Certain other constraints such as impedance behavior and physical size will have to be met. This antenna problem may be classed as a synthesis problem. Although some synthesis techniques will be considered, the approach in this text is to look at some elementary antennas, antenna systems, and reflectors which may be used as guides in obtaining the required characteristic.

5-3. ELEMENTAL DIPOLE ANTENNA

Radiation properties of many antennas may be obtained by assuming that the current distribution on the antenna consists of a superposition of current elements of appropriate magnitude and, where necessary, of appropriate phase. Knowing the fields produced by a current element, by superposition it is then possible to obtain the composite fields of the antenna. Of course, care must be taken to add the fields with proper consideration for their vector and phasor properties. As will be indicated later, an equivalent current source may also be defined for aperture-type antennas as well.

For these reasons and because additional characteristics of antennas may be illustrated by the simple current element, the elemental dipole is a good starting point in the study of antennas.

Vector and Scalar Potentials

The reader will perhaps recall that in the time-stationary cases, that is, electrostatics and magnetostatics, the electric and magnetic fields were obtained from a set of potentials, as follows:

Electrostatics

$$\mathbf{E} = -\nabla \Phi \tag{1}$$

$$\Phi(x,y,z) = \frac{1}{4\pi\epsilon} \int_V \frac{\rho(x',y',z')}{r} \, dx' \, dy' \, dz' \tag{2}$$

$$r = [(x-x')^2 + (y-y')^2 + (z-z')^2]^{\frac{1}{2}} \tag{3}$$

Magnetostatics

$$\mathbf{B} = \nabla \times \mathbf{A} \qquad \mathbf{H} = \frac{\mathbf{B}}{\mu} \tag{4}$$

$$\mathbf{A}(x,y,z) = \frac{\mu}{4\pi} \int \frac{\mathbf{i}(x',y',z')}{r} \, dx' \, dy' \, dz' \tag{5}$$

In the time-stationary regime there is no interdependence between the electric and magnetic fields nor is there any interdependence between the scalar and vector potentials.

When we come to time-varying fields, it is found that the electric and magnetic components of the field can again be constructed from scalar and vector potentials. In fact the potentials are related to those of the time-stationary fields, except that allowance must be made for the propagation of effects from the sources to the point of observation, as we saw in Chap. 2. The values of the scalar and vector potentials at a point P (Fig. 5-2) at the instant t are those given by the expressions above in terms of the values of the charges and currents at the source point (x',y',z') at the earlier time t',

$$t' = t - \frac{r}{v}$$

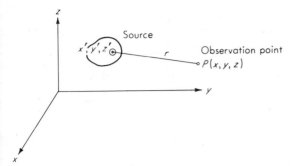

Fig. 5-2 Geometry relative to the computation of the potentials.

where v is the phase velocity in the medium. We write accordingly

$$\Phi(x,y,z,t) = \frac{1}{4\pi} \int \frac{[\rho]}{r} dV' \tag{6}$$

$$\mathbf{A}(x,y,z,t) = \frac{\mu}{4\pi} \int \frac{[\mathbf{i}]}{r} dV' \tag{7}$$

with the understanding that the bracketed sources mean that we use the values of the sources corresponding to the time t' as defined above.

The magnetic field is obtained from the vector potential $\mathbf{A}$ by the same process as used in the case of the static field, namely,

$$\mathbf{B} = \nabla \times \mathbf{A} \qquad \mathbf{H} = \frac{1}{\mu} \nabla \times \mathbf{A} \tag{8}$$

However, the electric field is developed out of both potentials, and the result is

$$\mathbf{E} = -\nabla \Phi - \frac{\partial \mathbf{A}}{\partial t} \tag{9}$$

This last equation shows explicitly the interrelationship between the electric and magnetic field that is characteristic of electromagnetic phenomena.

The vector and scalar potentials are related to one another by an equation that is similar to the equation of continuity. The relation is, in fact, an alternate statement of the equation of continuity. Thus

$$\nabla \cdot \mathbf{A} + \mu\epsilon \frac{\partial \Phi}{\partial t} = 0 \tag{10}$$

For harmonically time-varying fields this reduces to

$$\nabla \cdot \mathbf{A} + j\omega\mu\epsilon\Phi = 0 \tag{11}$$

where again the quantities are the complex amplitudes that are functions of position. Thus for the harmonic case the scalar potential can be computed directly from the vector potential, and we need compute only the latter from the sources according to (7).

In the case of the harmonically time-varying field, the time dependence throughout is of the basic form $e^{j\omega t}$. Thus our current is

$\mathbf{i}(x,y,z)e^{j\omega t}$

at the instant t. However, when we wish to determine the field at the point P at instant t, we must use the current at the source at instant t'; that is, the retarded value of the current at the source is

$$[\mathbf{i}] = \mathbf{i}(x',y',z')e^{j\omega t'} = \mathbf{i}(x',y',z')e^{j\omega(t-r/v)}$$
$$= \mathbf{i}(x',y',z')e^{j(\omega t - kr)} \qquad k = \frac{\omega}{v} = \frac{2\pi}{\lambda} \tag{12}$$

The vector potential now becomes

$$\mathbf{A} = \frac{\mu}{4\pi}\int \mathbf{i}(x',y',z')\frac{e^{-jkr}}{r}\,dV'e^{j\omega t} \tag{13}$$

and the expressions for the electric and magnetic fields are

$$\mathbf{E} = -\frac{1}{j\omega\mu\epsilon}\nabla(\nabla \cdot \mathbf{A}) - j\omega\mathbf{A} \tag{14}$$

$$\mathbf{H} = \frac{1}{\mu}\nabla \times \mathbf{A} \tag{15}$$

obtained by making use of (9) in conjunction with (11). We should keep in mind that the derivatives symbolically represented by the operators are with respect to the coordinates of the point of observation $P(x,y,z)$.

For the case of the elemental dipole, let us choose a current element of length l oriented so that its axis coincides with the z axis (Fig. 5-3). If we consider any infinitesimal section of length $dz\,\mathbf{a}_z$, we observe that

$\mathbf{i}\,dV = (i\,dS)\,dz'\,\mathbf{a}_z$

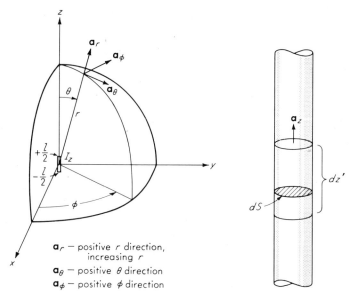

Fig. 5-3 Elemental dipole geometry.

a_r — positive r direction, increasing r
a_θ — positive θ direction
a_ϕ — positive ϕ direction

The vector potential has in this case only one component and is

$$A_z = \frac{\mu}{4\pi} \int_{-l/2}^{+l/2} I(z') \frac{e^{-jkr}}{r} dz' \qquad (16)$$

where dS is the cross-sectional area of the wire at the point under consideration and $\mathbf{a}_z$ is a unit vector along the wire. $i\,dS$ is the total current I flowing along the wire. The value may be a function of position along the wire; for the elemental dipole, I is assumed constant along the wire. For the elemental dipole, assume $l \ll \lambda$. Then r and $I(z')$ are constant over the integration, resulting in

$$A_z(r) = \mu \frac{Il}{4\pi r} e^{-jkr}$$

Using spherical coordinates, we write

$$A_r = A_z \cos\theta = \mu \frac{Il}{4\pi r} e^{-jkr} \cos\theta$$

and

$$A_\theta = -A_z \sin\theta = -\mu \frac{Il}{4\pi r} e^{-jkr} \sin\theta$$

from which, with (14) and (15), we obtain the following field components for the elemental dipole:

$$H_\phi = \frac{Il}{4\pi} e^{-jkr} \left(\frac{jk}{r} + \frac{1}{r^2}\right) \sin\theta = \frac{Il}{4\pi} e^{-jkr} k^2 \left(\frac{j}{kr} + \frac{1}{k^2 r^2}\right) \sin\theta$$

which may also be written as

$$\mathbf{H} = \mathbf{a}_\phi \left[\frac{j\omega}{4\pi}\left(\frac{jk}{r} + \frac{1}{r^2}\right) e^{-jkr} \sin\theta\right] \frac{Il}{j\omega}$$

$$E_r = \frac{Il}{4\pi} e^{-jkr} \left(\frac{2\eta}{r^2} + \frac{2}{j\omega\epsilon r^3}\right) \cos\theta = \frac{Il}{4\pi} e^{-jkr} 2\eta k^2 \left(\frac{1}{k^2 r^2} + \frac{1}{jk^3 r^3}\right) \cos\theta$$

and (17)

$$E_\theta = \frac{Il}{4\pi} e^{-jkr} \left(\frac{j\omega\mu}{r} + \frac{1}{j\omega\epsilon r^3} + \frac{\eta}{r^2}\right) \sin\theta$$

$$= \frac{Il}{4\pi} e^{-jkr} k\omega\mu \left(\frac{j}{kr} + \frac{1}{jk^3 r^3} + \frac{1}{k^2 r^2}\right) \sin\theta$$

which may in turn be expressed as follows:

$$\mathbf{E} = \left[\frac{1}{2\pi\epsilon}\left(\frac{1}{r^3} + \frac{jk}{r^2}\right) e^{-jkr} \cos\theta \, \mathbf{a}_r + \frac{1}{4\pi\epsilon}\left(\frac{1}{r^3} + \frac{jk}{r^2} - \frac{k^2}{r}\right) e^{-jkr} \sin\theta \, \mathbf{a}_\theta \right] \frac{Il}{j\omega}$$
(18)

where $\eta = (\mu/\epsilon)^{1/2} \approx 120\pi$ and $k = \omega(\mu\epsilon)^{1/2}$.

Far-zone or Radiation Field

An essential function of the antenna is to produce a distant field. We are therefore interested in the form of the fields for distances given by $2\pi r/\lambda = kr \gg 1$, remembering that the length of the antenna is also very small compared with the wavelength. For this case, the electric and magnetic fields for the elemental dipole reduce to the following:

$$E_\theta = j\omega\mu \frac{I_0 l}{4\pi r} \sin\theta \, e^{-jkr} \tag{19}$$

$$H_\phi = jk \frac{I_0 l}{4\pi r} \sin\theta \, e^{-jkr} = \left(\frac{\epsilon}{\mu}\right)^{1/2} E_\theta \tag{20}$$

So far we have directed attention to the way the amplitudes of the field components vary with position. Now let us look at the complex exponential, and see what it means. Let us consider a fixed point r and write

$$t' = t - \frac{r}{v}$$

Then we see that at instant t we are observing the field that would exist around a dipole at t' if there were no propagation effects. This is the retardation effect associated with wave propagation. The effect at a given instant t at a point is that of a source system at an earlier time corresponding to the interval required for the disturbance to reach the point of observation.

The complex exponential

$$e^{j(\omega t - kr)}$$

contains the representation of wave propagation in another sense. The complex exponential represents oscillation in time at any given point r. If we take two neighboring points r and $r + dr$, the field components will not have the same values because the complete expressions

$$\omega t - kr$$

and

$$\omega(t + dt) - k(r + dr)$$

represent different angles. However, at a time $t + dt$, the angle for the second point will be what it had been for the first point, if dt is the interval such that

$$\omega\, dt - k\, dr = 0$$

In other words, if we should pick a point in the field at instant t and move radially with a speed

$$v_p = \frac{dr}{dt} = \frac{\omega}{k}$$

we would observe no oscillation in the field intensity (only the change corresponding to the variation of the amplitude with r). That is just what we mean by propagation of the field disturbance and the wave velocity.

We note in the above discussion of the field of a dipole antenna that at large distances from the antenna the dominant term in the expression for the electric and magnetic fields is that varying as $1/r$. This part of the field, which we call the far-zone, or radiation, field, has a relatively simple structure in which the electric and magnetic field vectors are perpendicular to one another and in time phase with one another. This result for the elementary antenna is more general in scope and applies to any current distribution.

EXERCISES

5-1 Derive the expressions for the electric and magnetic fields (17) and (18).

5-2 Give a detailed derivation of the radiation from an infinitesimal dipole antenna of height l, carrying a uniform current $I_0 \exp(j\omega t)$. Use complex notation throughout; however, when you have completed the derivation, write the total field as is necessary if you wish to evaluate it at any given instant of time t. Fill in all steps of the derivation which were omitted above, and clearly state all assumptions which you make in the derivation.

5-3 A wire 1 m long is made to carry a *uniform* current of 10 amp at a frequency of 2 Mc.

a Calculate the electric field strength at a distance of 100 km in a direction at right angles to the axis of the wire.

b *Sketch* the radiation pattern, i.e., the field intensity at 100 km as a function of angle θ.

5-4. ENERGY FLOW AND POWER PATTERNS

As we saw in Chap. 2, the flow of energy in an electromagnetic field is represented by the Poynting vector,

P = **E** × **H**   watts/m²   (21)

The Poynting vector is the expression for the intensity of power flow; that is, if we have an element of area dS and the unit vector normal to it is **n**, the power flowing across a closed surface is

$W = \oint \mathbf{P} \cdot \mathbf{n}\, dS$ (22)

When we are dealing with fields whose time dependence is harmonic, we are interested in the average value of the power flow over a cycle rather than the instantaneous value. In this case we can use the complex representations for the field vectors. It turns out that the time-average Poynting vector is given by

$$\langle \mathbf{P} \rangle = \tfrac{1}{2} \operatorname{Re} (\mathbf{E} \times \mathbf{H}^*) \tag{23}$$

where the asterisk signifies that the complex conjugate is to be used. The average flow of power across a surface is then given by

$$\langle W \rangle = \oint \langle \mathbf{P} \rangle \cdot \mathbf{n}\, dS = \tfrac{1}{2} \operatorname{Re} \oint (\mathbf{E} \times \mathbf{H}^*) \cdot \mathbf{n}\, dS \tag{24}$$

As an example, let us compute the power flow from a dipole antenna by computing the value of W across a sphere of radius r surrounding the antenna. Referring to (17) and (18), we observe that

$$\mathbf{E} = \left[\frac{1}{2\pi\epsilon} \left(\frac{1}{r^3} + \frac{jk}{r^2} \right) \cos\theta\, \mathbf{a}_r + \frac{1}{4\pi\epsilon} \left(\frac{1}{r^3} + \frac{jk}{r^2} - \frac{k^2}{r} \right) \sin\theta\, \mathbf{a}_\theta \right] \frac{Il}{j\omega} e^{j(\omega t - kr)}$$

$$\mathbf{H}^* = \left[-\frac{j\omega}{4\pi} \left(\frac{1}{r^2} - \frac{jk}{r} \right) \sin\theta\, \mathbf{a}_\phi \right] \frac{I^*l}{-j\omega} e^{-j(\omega t - kr)}$$

The vector $\mathbf{E} \times \mathbf{H}^*$ will thus have a component in the θ direction and a component in the r direction, corresponding to the relations

$\mathbf{a}_r \times \mathbf{a}_\phi = -\mathbf{a}_\theta$
$\mathbf{a}_\theta \times \mathbf{a}_\phi = \mathbf{a}_r$

Note that the product of the exponential factors is just unity. When we take the real part of the complex vector product, we find that it is in the radial direction and is constructed from only the $1/r$ terms in the field components, as follows:

$$\langle \mathbf{P} \rangle = \frac{1}{2} \operatorname{Re} (\mathbf{E} \times \mathbf{H}^*) = \frac{k^3 \omega}{32\pi^2 \epsilon r^2} \sin^2\theta \, \frac{|I|^2 l^2}{\omega^2} \mathbf{a}_r \tag{25}$$

and the average flow of power out from the dipole is

$$\langle W \rangle = \oint \langle \mathbf{P} \rangle \cdot \mathbf{a}_r\, dS = \frac{k^3 \omega (|I|^2/\omega^2) l^2}{32\pi^2 \epsilon r^2} \int_0^\pi \int_0^{2\pi} \sin^2\theta\, r \sin\theta\, d\phi\, r\, d\theta$$

$$= \frac{l^2 k^3 \omega |I|^2 / \omega^2}{12\pi\epsilon} \tag{26}$$

In the development of the special case of the dipole antenna, we observe some general results. The average value of the Poynting vector is directed radially outward, corresponding to the sense of propagation of the wave from its source, and arises entirely from the radiation-field components. The average intensity of power flow varies inversely as the square of the distance from the source, so that the mean total power flowing out across a sphere surrounding the source is independent of the radius. This result corresponds to what must be the physical situation. If there is no loss in the medium, the power flowing across any one surface completely surrounding the source must be equal to the power flowing across any other surface completely surrounding the source.

Because of the above fundamental feature of the power flow, it is convenient and customary to describe the flow in terms of power per unit solid angle. Since the solid angle subtended by an element dS which is normal to the radius r is

$$d\Omega = \frac{dS}{r^2} \tag{27}$$

the power flow per unit solid angle in the direction of the solid angle $d\Omega$ is

$$W(\theta,\phi)\, d\Omega = r^2 \langle \mathbf{P} \rangle \cdot \mathbf{a}_r\, d\Omega \tag{28}$$

or

$$W(\theta,\phi) = r^2 \langle \mathbf{P} \rangle \cdot \mathbf{a}_r \tag{29}$$

The power per unit solid angle $W(\theta,\phi)$ is, in general, a function of the direction of observation θ,ϕ. In the case of the dipole, because of the symmetrical structure of the field about the dipole axis, the power flow is independent of the azimuth angle about the dipole axis, but the flow pattern is directive in character in the meridional planes. The distribution is given by the $\sin^2 \theta$ dependence.

Power patterns such as the dipole field pattern are sometimes called omnidirectional patterns. However, the description is misleading since the radiation pattern has directivity and has a null in the direction of the dipole axis.

We can now write down the corresponding results for a general current distribution by making use of the far-zone field expressions, which for this case can be shown to consist of only the following components (see Fig. 5-4):

$$\mathbf{E} = E_\theta \mathbf{a}_\theta + E_\phi \mathbf{a}_\phi \tag{30a}$$

$$\mathbf{H}^* = -\left(\frac{\epsilon}{\mu}\right)^{\frac{1}{2}} E_\phi^* \mathbf{a}_\theta + \left(\frac{\epsilon}{\mu}\right)^{\frac{1}{2}} E_\theta^* \mathbf{a}_\phi \tag{30b}$$

and as a result,

$$\langle \mathbf{P} \rangle = \frac{1}{2}\left(\frac{\epsilon}{\mu}\right)^{\frac{1}{2}}(|E_\theta|^2 + |E_\phi|^2)\mathbf{a}_r \tag{31}$$

The power per unit solid angle in the direction given by θ,ϕ is then

$$W(\theta,\phi) = r^2 \frac{1}{2}\left(\frac{\epsilon}{\mu}\right)^{\frac{1}{2}}(|E_\theta|^2 + |E_\phi|^2) \tag{32}$$

Since E_θ and E_ϕ vary inversely with r, $W(\theta,\phi)$ is independent of r.

The directivity of the power flow is specified in terms of a quantity known as the gain function. A hypothetical isotropic radiator radiating the same total power as the antenna under consideration would have a

Fig. 5-4 Far-zone field components.

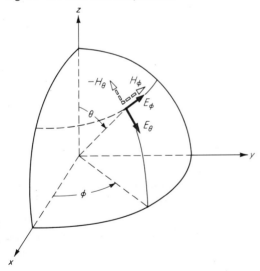

power pattern, or power per unit solid angle,

$$\frac{W_T}{4\pi}$$

The ratio of the actual power per unit solid angle in the direction θ,ϕ to the power per unit solid angle from the isotropic radiator is the gain function of the antenna in that direction,

$$g(\theta,\phi) = \frac{W(\theta,\phi)}{W_T/4\pi} \tag{33}$$

Now the total power radiated is simply the integral of the power per unit solid angle over the entire space as follows:

$$W_T = \int W(\theta,\phi)\, d\Omega = \int_0^\pi \int_0^{2\pi} W(\theta,\phi) \sin\theta\, d\phi\, d\theta \tag{34}$$

and the gain function is, therefore,

$$g(\theta,\phi) = \frac{4\pi W(\theta,\phi)}{\int_0^\pi \int_0^{2\pi} W(\theta,\phi) \sin\theta\, d\phi\, d\theta} \tag{35}$$

The maximum value of the gain function is known as the directivity. We usually express the directivity in decibels above the isotropic radiator as follows:

$$G = 10 \log g_{\max} \tag{36}$$

5-5. WIRE ANTENNAS

We shall consider several important cases and illustrate some of the procedures for calculating the radiation pattern. Generally one proceeds to do the calculations with the assumption that the current distribution is known. Actually, however, the nature of the physical problem is such that the distribution is not known. When one turns on a transmitter, he knows the configuration of the transmission line leading to the antenna and the structure of the antenna and he may make a measurement of the voltage and current at the input terminals of the transmission line. The real antenna problem, in the sense of a problem of electromagnetic theory, is to determine the current distribution. It is a most difficult problem and, in general, is beyond solution. What one does is to make some guess as to the current distribution and proceed on that basis. The several exact solutions that have been obtained provide one with a background and experience that help in further guesses on more intricate systems.

In the case of a thin wire antenna without any type of loading structure, the current distribution is assumed to be sinusoidal. It must go to zero at the ends of the wire. The assumed current distribution is an excellent approximation for thin wire antennas; major deviations occur at the feeding points, where the current distribution is very sensitive to the details of the physical structure.

Field Calculations with the Aid of the Elemental Dipole

The current distribution along a wire antenna is equivalent to a linear distribution of dipoles of moment

$$\frac{I(l')}{j\omega} dl'$$

The total magnetic field is the result obtained by superposition of dipole fields. Thus for a wire antenna consisting of a straight wire along the z axis and extending from z_1 to z_2, the magnetic field is as follows:

$$\mathbf{H} = \int_{z_1}^{z_2} \mathbf{a}_\phi \left[\frac{j\omega}{4\pi} \left(\frac{jk}{r} + \frac{1}{r^2} \right) e^{-jkr} \sin\theta \right] \frac{I(z')}{j\omega} dz' \qquad (37)$$

Similarly, the electric field results from a superposition of elemental dipoles, yielding

$$\mathbf{E} = \int_{z_1}^{z_2} \left[\frac{1}{2\pi\epsilon} \left(\frac{jk}{r^2} + \frac{1}{r^3} \right) e^{-jkr} \cos\theta\, \mathbf{a}_r \right. \\ \left. + \frac{1}{4\pi\epsilon} \left(\frac{1}{r^3} + \frac{jk}{r^2} - \frac{k^2}{r} \right) e^{-jkr} \sin\theta\, \mathbf{a}_\theta \right] \frac{I(z')}{j\omega} dz' \qquad (38)$$

In applying the foregoing results, we should keep in mind that the angle θ and the unit vectors inside the integral as well as the distance r vary with the source position. The integrals give the field everywhere, both at large distances from the wire and at points arbitrarily close to the wire. They are unfortunately very difficult to evaluate for an arbitrary field point; it is frequently necessary to resort to numerical methods of integration.

The result that we arrived at, namely, the current distribution can be regarded as a system of dipoles and the field of the distribution can be obtained by superposition of dipole fields, is actually of greater generality. An extended current distribution can be dealt with in a similar manner by considering each element $i\,dV$ as an elementary dipole having a dipole moment $i\,dV/j\omega$ (see Fig. 5-5) as follows:

$$\mathbf{E} = \int_V \left[\frac{1}{2\pi\epsilon} \left(\frac{jk}{r^2} + \frac{1}{r^3} \right) e^{-jkr} \cos\theta\, \mathbf{a}_r \right.$$
$$\left. + \frac{1}{4\pi\epsilon} \left(\frac{1}{r^3} + \frac{jk}{r^2} - \frac{k^2}{r} \right) e^{-jkr} \sin\theta\, \mathbf{a}_\theta \right] \frac{i\,dV}{j\omega} \quad (39)$$

$$\mathbf{H} = \int_V \frac{j\omega}{4\pi} \left(\frac{jk}{r} + \frac{1}{r^2} \right) (\mathbf{a}_i \times \mathbf{a}_r) e^{-jkr} \frac{i\,dV}{j\omega} \quad (40)$$

We shall not use the above expressions in detail to compute the fields of antennas, but it is important to understand the basic ideas contained in the expressions. The total field is the resultant of dipole fields, and once we have a knowledge of the dipole field, we can construct a general field by addition of vectors. The integration can be carried out graphically, which is often done in practice.

Fig. 5-5 Extended current distribution.

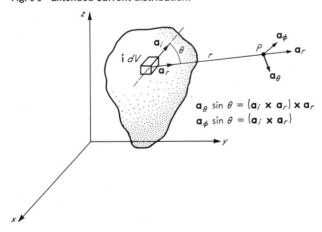

Illustrative Examples

We shall now apply the relationship for the electric field due to an elemental dipole to the calculation of the far-zone fields of several wire antennas.

Linear antenna. Consider first the linear antenna illustrated in Fig. 5-6a. The antenna is center-driven and has a length comparable to the wavelength. In order to proceed with the calculation for the arbitrary length of the antenna of Fig. 5-6b, it is necessary to have an expression for the current distribution. For this case it is adequate to assume the sinusoidal distribution discussed above, so that

$$I(z') = I_m \sin\left[\frac{2\pi}{\lambda}\left(\frac{l}{2} - z'\right)\right] \quad \text{for } z' > 0$$

and

$$I(z') = I_m \sin\left[\frac{2\pi}{\lambda}\left(\frac{l}{2} + z'\right)\right] \quad \text{for } z' < 0 \quad (41)$$

In the following expression for the far-zone electric field

$$E_\theta = \int_{-l/2}^{+l/2} \frac{1}{4\pi\epsilon} \frac{-k^2}{r'} e^{-jkr'} \sin\theta' \frac{I(z')}{j\omega} dz' \quad (42a)$$

Fig. 5-6 Linear antennas. (a) Full-wave antenna, $kl = 2\pi$; (b) arbitrary length antenna; (c) half-wave antenna, $kl = \pi$.

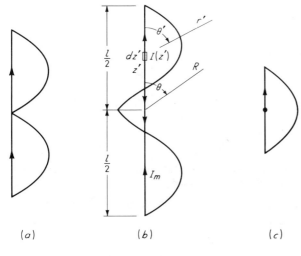

Approximations may be made if the point at which the field is to be determined is distant. Thus for $\theta \approx \theta'$

$$r' = (R^2 + z'^2 - 2Rz' \cos \theta)^{\frac{1}{2}}$$
$$\approx R - z' \cos \theta$$

and $1/r \approx 1/R$ in the amplitude variation of the fields, whereas in the phase factor

$$e^{-jkr'} = e^{-jkR+jkz'\cos\theta}$$
$$= e^{-jkR}e^{+jkz'\cos\theta}$$

since $kz' \cos \theta = (2\pi/\lambda)z' \cos \theta$ may be an appreciable number of radians.

Upon substitution of these approximations, the far-zone fields become

$$E_\theta = \eta H_\phi = j\eta \frac{kI_m}{4\pi R} \sin \theta \, e^{-jkR} \left\{ \int_{-l/2}^{0} e^{jkz'\cos\theta} \sin\left[k\left(\frac{l}{2} + z'\right) \right] dz' \right.$$
$$\left. + \int_{0}^{l/2} e^{jkz'\cos\theta} \sin\left[k\left(\frac{l}{2} - z'\right) \right] dz' \right\}$$
$$= j\eta \frac{kI_m}{4\pi R} \sin \theta \, e^{-jkR} \left\{ \frac{2}{k \sin \theta} \left[\cos\left(\frac{kl}{2} \cos \theta\right) - \cos \frac{kl}{2} \right] \right\}$$
$$= j\eta \frac{kI_m}{2\pi R} e^{-jkR} \left\{ \frac{\cos\left[(kl/2) \cos \theta\right] - \cos(kl/2)}{\sin \theta} \right\} \quad k = \frac{2\pi}{\lambda} \quad (42b)$$
$$\eta = (\mu/\epsilon)^{\frac{1}{2}}$$

For the *half-wave antenna*, the length $l = \lambda/2$, and since

$$\frac{kl}{2} = \left(\frac{2\pi}{\lambda}\right)\left(\frac{\lambda}{4}\right) = \frac{\pi}{2}$$

the fields become

$$E_\theta = \eta H_\phi = j\eta \frac{I_m}{2\pi R} e^{-jkR} \left\{ \frac{\cos\left[(\pi/2) \cos \theta\right]}{\sin \theta} \right\} \quad (43)$$

The resultant field is completely symmetrical about the z axis. Its directional characteristic (radiation pattern) is somewhat similar to that of a dipole antenna, differing in that it is somewhat more directive. The power gain of the half-wave antenna is 1.64, whereas that of the infinitesimal dipole is 1.5.

For the case of the *center-fed full-wave* antenna, $l = \lambda$ and

$$\frac{kl}{2} = \frac{2\pi}{\lambda}\frac{\lambda}{2} = \pi$$

for this length. Thus, the expression for the electric field is

$$\begin{aligned}E_\theta = \eta H_\phi &= j\eta \frac{I_m}{2\pi R} e^{-jkR}\left[\frac{\cos(\pi\cos\theta)+1}{\sin\theta}\right] \\ &= j\eta \frac{I_m}{2\pi R} e^{-jkR} 2\left\{\frac{\cos[(\pi/2)\cos\theta]}{\sin\theta}\right\}\cos\left(\frac{\pi}{2}\cos\theta\right) \end{aligned} \quad (44)$$

It is interesting to note that the system can be regarded as two half-wave antennas excited in phase, their centers being at a distance of $\lambda/2$ from one another. The resulting pattern is more directive than that of a single half-wave radiator (gain = 2.5). The maximum power radiated is in the direction $\theta = \pi/2$, the horizontal plane.

Circular Loop Antenna in Its Lowest Mode, Uniform Current

Figure 5-7 shows a circular loop antenna carrying a current that is constant in magnitude around the loop. This is the distribution that will be realized when the diameter of the loop is very small compared with the

Fig. 5-7 Circular loop antenna.

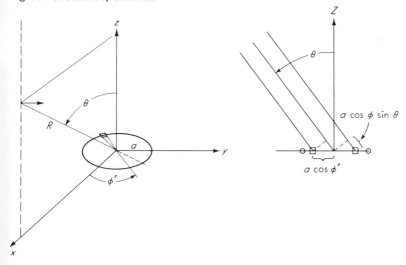

wavelength. Although the current is constant in magnitude, it is not constant in direction.

One can simplify the arithmetic involved by giving consideration to the symmetry of the configuration. The field must be independent of the azimuth angle about the z axis. We can, therefore, fix our attention at a point in the plane $\phi = 0$, that is, the xz plane, and consider how the dipole elements into which we can resolve the loop will interact at the point indicated. If we take the four symmetrically located points in the four quadrants of the loop, we can see that the resultant of their E_θ contributions is zero and that we have only an E_ϕ component in the resultant field. This component is

$$E_\phi = -\int_0^\pi \frac{j\omega\mu I_0 a \, d\phi' \cos \phi'}{4\pi R} \left(-e^{-jk(R+a\cos\phi'\sin\theta)} + e^{-jk(R-a\cos\phi'\sin\theta)} \right)$$

$$= \frac{j\omega\mu I_0 a}{4\pi R} e^{-jkR}(-2j) \int_0^\pi \cos \phi' \left(\frac{e^{-jka\cos\phi'\sin\theta} - e^{+jka\cos\phi'\sin\theta}}{2j} \right) d\phi'$$

$$= \frac{2\omega\mu I_0 a}{4\pi R} e^{-jkR} \int_0^\pi \sin(ka\cos\phi'\sin\theta) \cos \phi' \, d\phi'$$

Now $\int_0^\pi \sin(ka\cos\phi'\sin\theta)\cos\phi' \, d\phi' = \pi J_1(ka\sin\theta)$, where J_1 is a Bessel function of the first kind, order 1. Therefore, the fields for the circular loop become

$$E_\phi = -\eta H_\theta = \frac{\omega\mu I_0 a}{2R} e^{-jkR} J_1(ka\sin\theta) \tag{45}$$

For a very small loop for which $ka = 2\pi a/\lambda \ll 1$,

$$E_\phi = \frac{\omega\mu I_0 a}{2R} e^{-jkR} \frac{1}{2} ka \sin \theta$$

$$= \omega\mu I_0 \frac{a^2 k}{4R} e^{-jkR} \sin \theta \tag{46}$$

Figure 5-8 illustrates the field variation for the simple wire antennas considered. Obviously, only the simplest communication system's specifications can be met using these antennas of relatively low directivity.

EXERCISE

5-4 Outline and carry out as far as possible the detailed determination of the gain factor for the half-wave antenna, infinitesimal dipole antenna, and the infinitesimal loop antenna.

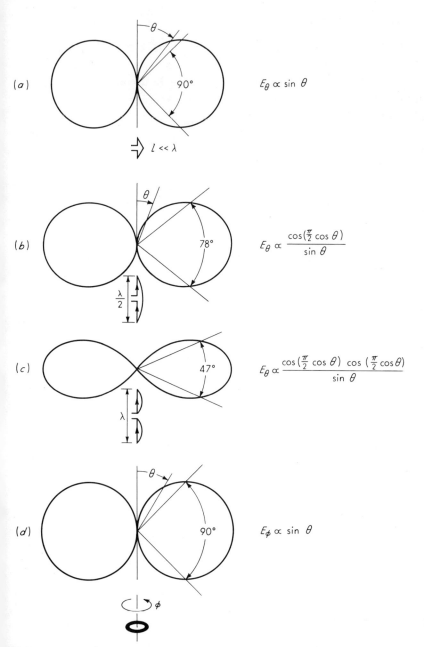

Fig. 5-8 Far-field radiation patterns of simple wire antennas. The half-power (0.707 field) angles are shown ranging from 90 to 47°. (a) Elemental dipole; (b) half-wave antenna; (c) full-wave antenna; (d) loop.

5-6. ANTENNA ARRAYS

The combination of systems of currents leads to a variety of patterns generally of greater directivity than the pattern of a dipole. The directivity comes about by the interference between the fields of the individual elements, resulting in an overall smaller field intensity, and by constructive interaction (in-phase superposition), resulting in enhancement of the field intensity. Power flow is proportional to the square of the field intensity, so that the power pattern is affected even more markedly.

A particularly important technique of obtaining increased directivity is that of purposely combining current distributions in a periodic arrangement. The resulting pattern is determined by the intrinsic phase relations between the members of the array and the spacing. We shall introduce the basic ideas by considering some simple particular cases.

The simplest system is the doublet, a two-element array. The elements can be a pair of elementary dipole antennas, a pair of half-wave antennas, or, indeed, any pair of similar antennas. Let us take the elements to be a pair of similar half-wave antennas, and by similarity we mean that they have the same orientations; it is possible to set one in correspondence with the other by a pure translation (no rotation of the structure is required). The system is shown in Fig. 5-9. As the figure indicates, we are going to consider the field in the far zone of the entire system. We are at distances such that $R \gg d$ and $R \gg l$, the length of an antenna. Then we certainly are in the far zone of the individual antennas, and in the far zone we may take the field to be just the radiation field (we want only the part of the total field that depends on R as $1/R$).

Fig. 5-9 A two-element array.

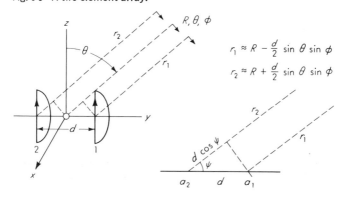

Now the field of antenna 1, referred to its center as the origin, is

$$E_{\theta_1} = j\left(\frac{\mu}{\epsilon}\right)^{\frac{1}{2}} \frac{I_{0_1}}{2\pi r_1} e^{-jkr_1} \frac{\cos\left[(\pi/2)\cos\theta\right]}{\sin\theta} \tag{47}$$

and that of antenna 2, referred to its center, is

$$E_{\theta_2} = j\left(\frac{\mu}{\epsilon}\right)^{\frac{1}{2}} \frac{I_{0_2}}{2\pi r_2} e^{-jkr_2} \frac{\cos\left[(\pi/2)\cos\theta\right]}{\sin\theta} \tag{48}$$

Since we are in the far zone, the θ directions associated with the two antennas individually differ from each other and from the θ direction of the general coordinate system with origin at 0 only in second order. Furthermore, we can introduce the far-zone approximations for r_1 and r_2 in terms of R, which we use in the phase factors. The resultant field is then

$$E_\theta = E_{\theta_1} + E_{\theta_2} = j\left(\frac{\mu}{\epsilon}\right)^{\frac{1}{2}} \frac{e^{-jkR}}{2\pi R} \frac{\cos\left[(\pi/2)\cos\theta\right]}{\sin\theta} (I_{0_1} e^{(jkd/2)\sin\theta\sin\phi}$$
$$+ I_{0_2} e^{(-jkd/2)\sin\theta\sin\phi}) \tag{49}$$

The two half-wave antennas are similar, but we can have a different level of current in one from the other and also a phase difference between them; that is, we can have the general condition

$$I_{0_2} = a_2 I_0 e^{jx} \qquad I_{0_1} = a_1 I_0 \tag{50}$$

The resultant field then takes the form

$$E_\theta = \left[j\left(\frac{\mu}{\epsilon}\right)^{\frac{1}{2}} \frac{I_0 e^{-jkR}}{2\pi R} \frac{\cos(\pi/2)\cos\theta}{\sin\theta}\right] \Lambda(\theta,\phi) \tag{51}$$

where

$$\Lambda(\theta,\phi) = a_1 e^{(j\pi d/\lambda)\sin\theta\sin\phi} + a_2 e^{jx} e^{(-j\pi d/\lambda)\sin\theta\sin\phi} \tag{52}$$

Before discussing any of the particular features of doublet arrays, let us take note of the form of (52), which carries over to more general arrays. We observe that the resultant pattern is the product of the pattern of a single element (the term in brackets) and a factor that depends only on the relative positions and phases of the reference points of the individual elements. We call this factor the array factor. If the individual elements were isotropic radiators, the array factor would be their resultant pattern with the phase reference point having been taken at the center. From the nature of the construction of the field, it should be evident how we can extend the method to a larger number of elements. It should also be noted that the array factor represents the addition of wavelets from the separate elements; the total phase of the contribution of an element to the field is determined by the intrinsic phase of the source (with respect to some reference, of course) and by the phase delay represented by path length from the source to the point of observation. Only relative phases are significant in determining the resulting power pattern, and we may take any element or other convenient point as the origin for measuring relative path lengths.

Finally, we note that the pattern of the array (array factor alone) is symmetrical about the axis of the array. We used the coordinate system shown in Fig. 5-9 because the field of the individual elements is expressed most simply in that coordinate system. However, insofar as the array is concerned, it would have been better to use the angle ψ shown in the figure. We would have obtained for the array factor the following:

$$\Lambda(\theta,\phi) \rightarrow \Lambda(\psi) = a_1 e^{(j\pi d/\lambda)\cos\psi} + a_2 e^{(-j\pi d/\lambda)\cos\psi} e^{jx}$$

Returning to the doublet system, we shall consider the case when the two elements have the same current levels; that is, $a_1 = a_2 = 1$. We can then combine the exponential terms in the array factor and obtain

$$E_\theta = j\left(\frac{\mu}{\epsilon}\right)^{\frac{1}{2}} \frac{I_0 e^{-jkR} e^{+j(x/2)}}{\pi R} \frac{\cos\left[(\pi/2)\cos\theta\right]}{\sin\theta} \cos\left(\frac{\pi d}{\lambda}\sin\theta\sin\phi - \frac{x}{2}\right) \quad (53a)$$

$$\Lambda(\psi) \rightarrow 2e^{j(x/2)} \cos\left(\frac{\pi d}{\lambda}\sin\theta\sin\phi - \frac{x}{2}\right)$$

$$= 2e^{j(x/2)} \cos\left(\frac{\pi d}{\lambda}\cos\psi - \frac{x}{2}\right) \quad (53b)$$

An important case is that when the spacing d is a quarter wavelength and the phase angle x is $\pi/2$. Then the array factor becomes

$$\Lambda(\psi) = 2e^{j\pi/4} \cos\left(\frac{\pi}{4}\cos\psi - \frac{\pi}{4}\right) \quad (54)$$

and we see that in the $+y$ direction ($\psi = 0$) the array factor is equal to 2, whereas in the $-y$ direction ($\psi = \pi$) the array factor reduces to zero. Thus in one direction the resultant field intensity is twice that of a single radiator, whereas in the opposite direction the resultant is zero. The reader should try to construct this result from just a direct consideration of the phases of the wavelets and their dependence on the phases of the sources and the differences in path lengths. The reader is also left with the exercise of constructing the total pattern of the doublet which we have just treated and estimating the gain of the system.

Uniform Arrays

One important class of arrays is that in which the elements have equal levels of excitation and there is a progressive phase delay or phase difference from element to element. The excitation is given by

$$A = A_0 e^{-jnx_0} \tag{55}$$

where we are taking the origin as the phase reference. We shall in fact take the origin at one end of the array and count $n = 0, 1, 2, \ldots, N - 1$, where N is the total number of elements in the array. The array factor is then

$$\Lambda(\psi) = A_0 \sum_{n=0}^{N-1} e^{jn(kd \cos \psi - x_0)} = A_0 \sum_{n=0}^{N-1} (e^{j(kd \cos \psi - x_0)})^n \tag{56}$$

This is a geometric series of N terms and can be summed. The result, after a few simple steps, is found to be

$$\Lambda = \frac{1 - e^{jN(kd \cos \psi - x_0)}}{1 - e^{j(kd \cos \psi - x_0)}} = e^{j[(N-1)/2](kd \cos \psi - x_0)} \frac{\sin N[(\pi d/\lambda) \cos \psi - (x_0/2)]}{\sin [(\pi d/\lambda) \cos \psi - (x_0/2)]} \tag{57a}$$

and the absolute value of the array factor is

$$|\Lambda| = \left| \frac{\sin N[(\pi d/\lambda) \cos \psi - (x_0/2)]}{\sin [(\pi d/\lambda) \cos \psi - (x_0/2)]} \right| \tag{57b}$$

The part given by (57b) is the essential part since it determines the power pattern of the array. The form of the function is shown in Fig. 5-10. The larger the value of N, the greater the height of the major peaks

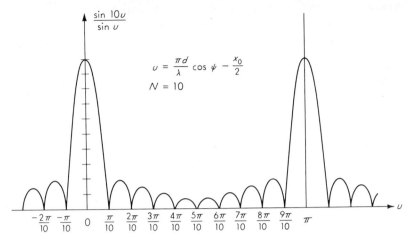

Fig. 5-10 A 10-element array.

above the side lobes and the narrower the main lobes of the function. The structure of the array pattern becomes the dominating feature of the pattern of the system. Thus, if the element pattern is not especially directive and, in particular, if the element pattern does not have a null in the direction corresponding to the major peak of the array pattern, the array pattern determines the direction of maximum intensity and also the side-lobe structure about the main lobe.

Note that the array factor is periodic in the variable u. However, the pattern of the physical system must be limited to the range of u that corresponds to $0 < \psi < \pi$, or

$$-\frac{\pi d}{\lambda} - \frac{x_0}{2} < u < \frac{\pi d}{\lambda} - \frac{x_0}{2} \tag{58}$$

We can thus obtain the radiation pattern of the array from the portion of the periodic function $\sin Nu / \sin u$ that is encompassed by the above range of u.

When the phase delay x_0 is zero, we get a principal maximum (main lobe) in the direction $\psi = \pi/2$; that is, we have a broadside array. When the spacing is a quarter wavelength, or $d = \lambda/4$, and the phase delay is $90°$, or $x_0 = \pi/2$, we find that the principal maximum occurs in the direction $\psi = 0$, that is, along the axis of the array; such an array is referred to as an end-fire array. For a given spacing, it is possible to shift the position of the principal maximum by varying the phase delay x_0.

We obtain then a scanning antenna which is particularly useful in radar applications.

The techniques for adjusting the amplitudes and phases of the elements of an array depend on the nature of the feeding system and the nature of the elements. For example, a set of half-wave antennas can be arranged as shunt elements on a two-wire line (see Fig. 5-11), or a set of dipoles can be excited from a waveguide, or the elements can be a set of slots in the wall of the waveguide. The phase of the elements is determined by the line length from element to element and also the nature of the loading. Details will be found in the various references on antennas.

Nonuniform Arrays

One of the drawbacks of the uniform array is the relatively high side lobes. Considerable research has been devoted to finding methods for reducing the side lobes. This reduction can be accomplished by varying the excitation along the array using a variety of methods. The general idea for the case of the broadside array is to have the excitation increase from one end of the array to the center and then decrease again to the other end. The most famous of the techniques are the so-called Tchebysheff arrays and the binomial arrays. The reduction in side-lobe level is obtained at the cost of widening the main lobe. It is necessary to make some compromises between the various operational requirements.

Fig. 5-11 Broadside arrays. (a) Shunt loading of a two-wire line with dipole elements; note reversal of the leads to compensate for the 180° phase difference represented by the length of line between elements; (b) slot array in a rectangular waveguide; staggering about the center line is equivalent to reversing the leads. The arrows indicate the current flow.

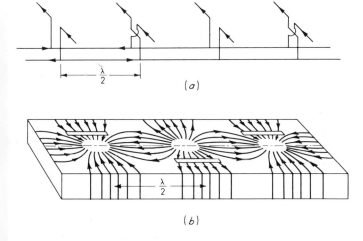

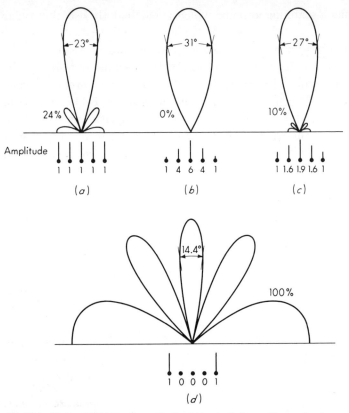

Fig. 5-12 Tapered illumination effect on the radiation pattern of arrays. (a) Uniform; (b) binomial; (c) optimum; (d) end-excited. The spacing of the elements is one-half wavelength.

As an illustration, consider a five-element array having all the currents in phase. The first array has equal amplitude of currents (Fig. 5-12a). This yields a "half-power" beamwidth of 23° and a side-lobe level 24 percent of the field amplitude of the main beam. When the amplitudes of the currents are adjusted according to the coefficients of the binomial expansion, the main beamwidth is 31° with no side lobes (Fig. 5-12b). Figure 5-12c illustrates an amplitude arrangement yielding 27° beamwidth and equal side-lobe level of 10 percent for all lobes. Figure 5-12d demonstrates extreme amplitudes. The end elements have large values, whereas the others have zero amplitudes. This case of "extreme untapering" results in a sharp "main beam" but 100 percent side-lobe level.

EXERCISES

5-5 An infinitesimal dipole antenna is located at the point 1 and has a current $Ie^{i\pi/4}$ flowing. Another antenna of identical structure located at point 2 has a current $Ie^{-i\pi/4}$ flowing.

a If the antennas are arranged so that their axes are perpendicular to the plane $ABCD$ and are $\lambda/4$ m apart, obtain an expression for the electric field radiation pattern in the plane $ABCD$.

b Sketch the radiation pattern in the plane $ABCD$.

c Repeat (a) and (b) for the case for which the antennas are oriented with axes *in* the plane $ABCD$ and parallel to the side AD.

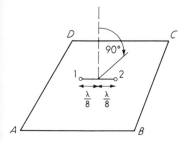

5-6 Ten vertical half-wave dipole antennas are spaced $\lambda/2$ apart and excited in phase with equal currents.

a Calculate all values of ϕ for the relative maxima and zeros of E_θ in the plane $\theta = \pi/2$, and then sketch a polar plot of E versus ϕ in this plane.

b How should the phases of the exciting currents be adjusted to transform this broadside array into an end-fire array (main lobe *along* line of array)?

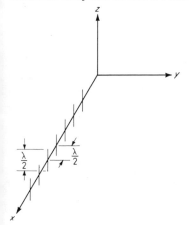

c What, if any, is the relationship between the radiation of the end-fire array and the broadside array in this case ($\theta = \pi/2$)?

d If θ is chosen differently than $\theta = \pi/2$, does your answer to **(c)** need modification?

5-7 Can unilateral directivity be achieved with a broadside array? (Unilateral directivity means that the main lobe is in a ϕ direction and there is no main lobe in the direction $\phi + \pi$.) Can it be achieved with an end-fire array? How?

5-8 Four identical antennas carrying the same current amplitude and phase are arranged symmetrically in a circle of radius $a + \lambda/4$. Assume that the antennas are infinitesimal dipoles placed perpendicular to the plane containing the circle.

a Find the variation of the far-zone field with θ measured in the plane of the circle.

b Sketch the pattern.

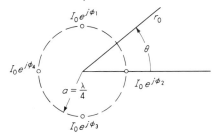

5-9 Suppose now in Exercise 5-8 that the magnitudes of the current elements are kept the same but that the phases of the currents are adjusted so that a maximum field is obtained at $\theta = 0$ and a zero field is obtained at $\theta = \pi$. What may be the phase angles ϕ_1, ϕ_2, ϕ_3, and ϕ_4?

Image Techniques

One technique for making an array is based on the principle of images and is accomplished by placing the antenna in the neighborhood of a large conducting sheet. For our purposes we shall consider the conducting sheet to be infinite in extent and perfectly conducting. The field of the antenna is reflected by the sheet, creating a field that corresponds to a second antenna located at the optical image position behind the sheet. The sense of the image is determined by the fundamental condition that at a perfectly conducting surface the tangential component of the total electric field must vanish. We need to determine the images for just the elementary dipole, and we can then construct the image for any other system by superposition.

Figure 5-13 shows the images for the two basic orientations of a dipole, parallel and perpendicular to the ground plane (ground plane is a term very often used for the image plane). Any other orientation of a

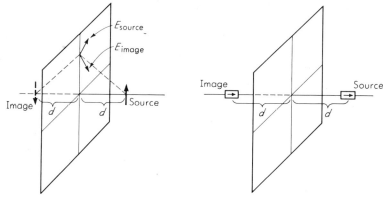

Fig. 5-13 Image sources.

dipole can be resolved into the two basic orientations. Observe that when the dipole is parallel to the ground plane, its image is 180° out of phase; when the dipole is normal to the plane, its image has a sense of being in phase. The reader should verify the relationship by setting up the fields at the boundary and studying the total tangential field.

The image technique can be used to develop a two-element array of dipoles which are out of phase by 180° for any desired spacing between the dipoles. This is more easily accomplished than by setting two dipole antennas up in space. The dipole–ground plane combination produces a pattern only in half the space; there is no field on the side of the image. Such an arrangement is particularly useful when it would be wasteful to transmit power in a region or when it is desirable to prevent reception from a region. For example, a TV antenna on the Farallon Islands off the coast from San Francisco would have little reason for having an ability to receive from the ocean side. Thus the image plane can be used to obtain a higher degree of directivity. A widely used type of directive antenna is made of a two-dimensional array of half-wave antennas arranged in a plane backed by an image plane. It is not necessary to use the ideal infinite plane; the ideal is approximated by a plane of finite extent that extends several wavelengths beyond the area of the dipole array. Furthermore, the plane can be perforated, provided the holes have dimensions small compared with the wavelength. For the region of several hundred megacycles, the plane can be replaced by a chicken-wire screen.

The plane screen is the simplest application of the image principle.

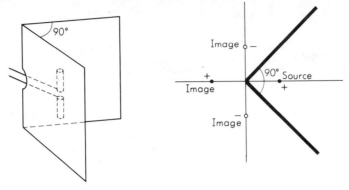

Fig. 5-14 Corner-reflector antenna.

Two-dimensional and even three-dimensional arrays can be formed. If a dipole antenna is placed between a pair of intersecting plane sheets, a directive system is obtained that is equivalent to an array along a circle. The method of developing the image system is the same as that used in electrostatics, taking into account the fundamental images shown in Fig. 5-13. Figure 5-14 shows the development of a 90° dihedral corner reflector. Here again it is not necessary to use infinite planes. A useful antenna can be made by using chicken-wire screens about four wavelengths on a side. The dipole is placed within a wavelength from the vertex; the distance can be varied to suit the particular application. Such an antenna has quite good characteristics and is one of the least costly directive antennas that can be constructed. Sets of dihedral antennas can be arranged to form an array, and the efficiency is remarkably good. Such antennas can be made economically for use over the VHF and UHF bands.

EXERCISE

5-10 A half-wave dipole antenna oriented parallel to the z axis is located at $(0,a,0)$. The plane $y = 0$ is a perfect conductor. The maximum current flowing in the antenna is I_0.
   a Find the radiated electric field $E(r,0,\phi)$.
   b Find the radiation resistance R_r. Does the conducting plane influence R_r?
   c Find the directive gain $g(\theta,\phi)$. Does $g(\theta,\phi)$ depend on any other parameter?
   d Evaluate the rms value of E for $I = 1$ amp, $r = 50$ km, $\theta = \pi/2$, and $\phi = \pi/2$, for $a = \lambda/2$, for $a = \lambda/4$, and for $a = 0$.
   e Repeat (d) for $\phi = 3\pi/2$.

5-7. APERTURE-TYPE ANTENNAS

Another important class of antennas is made up of reflectors, lenses, and horns in which there is an aperture that defines the transformation from guided waves to space waves. The corner-reflector system which we described briefly in the preceding section is, in a sense, such an antenna, since in the practical case the image planes are finite in extent and thus define an aperture, or, in the language of optics, an exit pupil.

Figure 5-15 shows a paraboloidal reflector antenna in which the primary radiator is a half-wave antenna backed by a small-plane reflector (using the image principle to make a directive source). The figure also shows a lens antenna with a similar primary radiator. The primary source is, in effect, a point source in most cases where such systems are employed. On the basis of ray optics, we can say that the optical device takes the rays diverging from the point source and converts them into a family of parallel rays. If this situation existed and the rays remained parallel to infinity, the far-zone beamwidth would be essentially zero and we would have an extremely directive system. Actually, however, the rays do not remain parallel, but the beam diffuses and develops a finite width; also, side lobes appear. The phenomenon is known as diffraction.

Diffraction is, of course, tied in with the fundamental mechanism of wave propagation. This mechanism was formulated many years ago in the study of physical optics by Huygens and Fresnel. The Huygens-Fresnel principle asserts that every point on a wavefront is a source of a secondary wavelet and the field at a subsequent point is the resultant of the contributions of the wavelets from all points on the previous wave surface. A system such as the paraboloid or lens generates a plane wave surface over its aperture. It is not a complete plane wave surface, which would be infinite, but instead is a delimited surface, according to the boundary of the aperture. Each point sends out spherical wavelets, and the resultant field in space arises by propagation and superposition of those wavelets. As a result the beam diffuses, and secondary maxima arise in those directions where the wavelets arrive in phase (or largely in phase as compared with neighboring directions).

An analysis of the problem of wave propagation results in identifying the equivalent Huygens sources to be a system of electric currents and magnetic currents. Note the term *equivalent*. Thus the field over the aperture is, in effect, a sheet of electric and magnetic currents or, what is

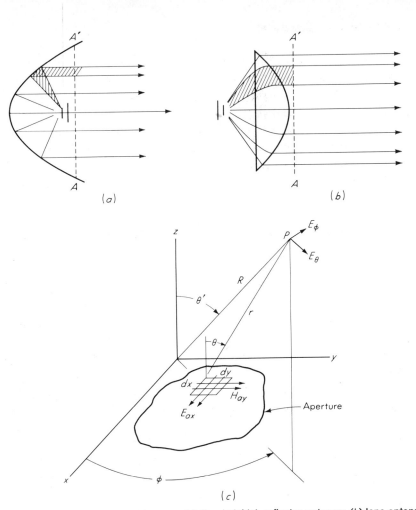

Fig. 5-15 Aperture-type antennas. (a) Paraboloidal reflector antenna; (b) lens antenna; (c) aperture geometry.

also equivalent thereto, a sheet of electric and magnetic dipoles. The problem of determining the properties of an aperture antenna is one of computing the field from a sheet of electric and magnetic dipoles. No small problem, of course, is that of determining the field over the aperture from which the equivalent sources are to be obtained.

An approximate solution can be obtained by the Huygens method for distant points P. It can be shown that the electric field components are given by

$$E_\theta = -j\frac{k}{4\pi}(E_{ax} + \eta H_{ay}\cos\theta)\cos\phi\,\frac{e^{-jkr}}{r}\,dx\,dy$$
$$E_\phi = +j\frac{k}{4\pi}(E_{ax}\cos\theta + \eta H_{ay})\sin\phi\,\frac{e^{-jkr}}{r}\,dx\,dy \quad (59)$$

where E_{ax} and H_{ay} are the field components in the aperture and where the angles and the geometry are as shown in Fig. 5-15c. In general, E_{ax} and H_{ay} (and, of course, other components, if they are present) are functions of position in the aperture and are not usually in time phase.

To illustrate the use of the relations of (59), consider the radiation pattern of a portion of a plane wavefront. For the plane wave traveling in the $+z$ direction, the elemental surface $dx\,dy$ has $E_{ax} = H_{ay}$ and the distant field E_θ is $1 + \cos\theta$ for $\phi = 0°$. The wavefront has a pattern directing the electromagnetic energy along $+z$; that is, for $\theta = 0°$ the field is maximum, whereas for $\theta = 180°$ all components are zero.

Next we consider a *uniformly illuminated rectangular aperture*. The aperture, illustrated in Fig. 5-16a, has an assumed electric and magnetic field that is uniform and in phase over the aperture. In calculating the distant field, we find the computations of the patterns are rather tedious. Integration of the contributions of each element must take into account the phase difference of the elements due to differences in distance to the far point.

Fig. 5-16 Uniformly illuminated apertures. (a) Rectangular; (b) circular.

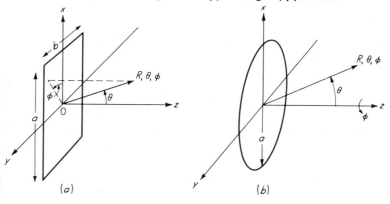

The electric field at P is

$$E_p = j\frac{k}{4\pi R} e^{-jkR} E_a(1 + \cos\theta)ab \frac{\sin[(\pi a/\lambda)\cos\theta_1]}{(\pi a/\lambda)\cos\theta_1} \frac{\sin[(\pi b/\lambda)\cos\theta_2]}{(\pi b/\lambda)\cos\theta_2} \tag{60}$$

where θ_1 is the angle between R and x and θ_2 is the angle between y and R. In the xz plane, $\theta_1 = 90° - \theta$ and $\theta_2 = 90°$, so that

$$E_p = j\frac{k}{4\pi R} e^{-jkR} E_a(1 + \cos\theta)ab \frac{\sin[(\pi a/\lambda)\sin\theta]}{(\pi a/\lambda)\sin\theta} \quad xz \text{ plane} \tag{61}$$

For $\theta = 0°$, the point P lies on the z axis and there the radiation has its maximum value

$$E_{\max} = \frac{k}{4\pi R} 2E_a ab = \frac{ab}{\lambda R} E_a \quad \text{along the } +z \text{ axis}$$

Since E_a and H_a are uniform and in phase over the aperture, the total average radiated power is

$$W = \tfrac{1}{2} E_a H_a ab = \frac{ab E_a^2}{240\pi}$$

The maximum gain of the uniformly illuminated rectangular aperture is therefore, by Eq. (33),

$$\begin{aligned} g_{\max} &= \frac{W(\theta,\phi)}{W_T/4\pi} = \frac{\tfrac{1}{2}(abE_a/\lambda R)(abE_a/120\pi\lambda R)R^2}{(abE_a^2/240\pi)/4\pi} \\ &= ab\frac{4\pi}{\lambda^2} = A\frac{4\pi}{\lambda^2} \end{aligned}$$

where $A = ab$ is the area of the uniformly illuminated rectangular aperture.

The shape of the beam produced by an aperture antenna depends on the shape of the aperture and the nature of the distribution of the sources over the aperture. The ideal situation from the standpoint of gain and beamwidth occurs when the distribution over the aperture is uniform in magnitude and phase. As we have seen, in that case, the gain of the aperture is

$$g = \frac{4\pi A}{\lambda^2} \qquad (62)$$

where A is the area of the aperture. This case of uniform illumination forms a reference system for the discussion of aperture antennas. It is often stated that the gain given by (62) is the maximum gain. There are, however, certain conditional factors which are of theoretical importance in the treatment of maximizing the gain, but they are not of great practical significance, at least not so far.

By tapering the intensity of the aperture distribution toward the edges, the gain is reduced and the beam is widened. In a sense, (62) is still effective, for by tapering the distribution we are effectively decreasing the area over which there is illumination. Tapering the illumination has one salutary effect, that of reducing the side-lobe intensities. Therefore, one has to compromise between gain and beam sharpness and side-lobe level.

In speaking of the gain for which we have set down (62), we were careful to refer to it as aperture gain. In the case of the paraboloid or lens systems of Fig. 5-15, it is clear that a fraction of the primary energy is not intercepted by the reflector or lens and, therefore, does not enter into the formation of the main beam. Hence if we refer to the overall gain defined by (33), it will be some fraction of the aperture gain. Thus the gain of an antenna is less than that of the uniformly illuminated aperture by virtue of spillover and tapering of the distribution. We ascribe to an aperture an effective area A_{eff} such that the gain is given by

$$g = \frac{4\pi A_{\text{eff}}}{\lambda^2} \qquad (63)$$

and in almost every practical case

$$\frac{A_{\text{eff}}}{A} < 1$$

Returning to the uniformly illuminated rectangular aperture, we note that the factor

$$(1 + \cos\theta)\frac{\sin[(\pi a/\lambda)\sin\theta]}{(\pi a/\lambda)\sin\theta}$$

except for the multiplier $1 + \cos \theta$, is very much like the array factor of Sec. 5-6. For a reasonably large ratio a/λ, the variation of $1 + \cos \theta$ with the angle θ is rather slow compared with the rest of the array factor.

In order to see how the beam structure is related to the aperture, we shall set down the most important terms in the field patterns of the rectangular aperture and a circular aperture; see Fig. 5-16 for the meaning of the quantities involved. The angles θ_1, θ_2, and θ_3 are written in terms of θ and ϕ. The power density patterns are as follows:

a. Rectangular aperture, uniformly illuminated

$$P(\theta,\phi) \propto \left\{ \frac{\sin\left[(\pi a/\lambda)\sin\theta\cos\phi\right]}{(\pi a/\lambda)\sin\theta\cos\phi} \frac{\sin\left[(\pi b/\lambda)\sin\theta\sin\phi\right]}{(\pi b/\lambda)\sin\theta\sin\phi} \right\}^2 \tag{64}$$

b. Circular aperture, uniformly illuminated

$$P(\theta,\phi) \propto \left\{ \frac{J_1[(\pi D/\lambda)\sin\theta]}{(\pi D/\lambda)\sin\theta} \right\}^2 \quad \begin{array}{l} D = \text{diameter} = 2a \\ J_1 = \text{Bessel function of order 1} \end{array} \tag{65}$$

In the case of the rectangular aperture, the pattern is seen to have the xz plane and the yz plane as planes of symmetry corresponding to the symmetry of the aperture. Each factor in (64) is of the basic form

$$\left(\frac{\sin u}{u}\right)^2$$

which is a function which we encountered in signal analysis as the spectrum of a rectangular pulse. Here the pulse is in the form of a distribution over the boundary plane $z = 0$, and the power pattern is a spatial spectrum corresponding to that distribution. The analogy which we are drawing here is actually a precise one that arises from the general theory. The student should study the $(\sin u)/u$ function again and refresh his memory on its general features. The width of the main lobe of the function, that is, the value of u for which the function has the value 0.707, is

$$u = \pm 1.39$$

The power function, which is the square of $(\sin u)/u$, has one-half the maximum value for the above value of u. On taking the pattern in each of the principal planes $\phi = 0$ and $\phi = \pi/2$, respectively, and making use of the value of u, we obtain the full half-power width of the main lobe in each principal plane. Thus

$$\Theta = 0.88 \frac{\lambda}{a} \qquad (66a)$$

$$\Theta = 0.88 \frac{\lambda}{b} \qquad (66b)$$

The intensity of the first side lobe, relative to the peak intensity, is 13.2 db below the peak power level.

We see that the beam is narrower in that principal plane where the aperture has the larger dimension. This is illustrative of another general feature of aperture systems: Increasing the aperture dimension relative to the wavelength results in a sharpening of the beam (Fig. 5-17). One advantage of going to shorter wavelengths is the increased directivity that can be obtained by the increased aperture-to-wavelength ratio.

The cross section of the beam from a rectangular aperture has an ovoidlike shape (considering just the main lobe). Very often an aperture is made so that the beam is very narrow in one plane and very broad in the other. Such a "beaver-tail" beam is combined with another at an angle to form a glide-path system or a height-finding system. If a symmetrically shaped narrow beam is required, we use a *circular aperture* with as symmetrical an aperture distribution as we can possibly get.

Fig. 5-17 Horn antennas. (a) Open-end waveguide; (b) E field flare.

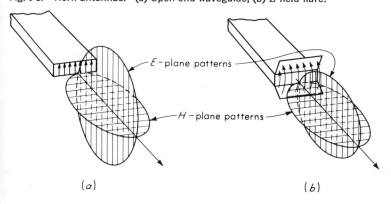

The pattern corresponding to (65) has a beamwidth

$$\theta = 1.02 \frac{\lambda}{D} \qquad (67)$$

and a first side lobe (power) of 17.6 db below the peak. Equation (67) again exhibits the fundamental principle that the beam sharpness varies directly with the ratio of dimensions to wavelength. Observe the difference between the side-lobe levels of the rectangular aperture and the circular aperture. Both cases are of uniform illumination. The difference is associated with the difference between the weightings of the distribution over the available aperture area.

The narrow, rotationally symmetric beam is used for search and tracking. It is nothing more than a searchlight beam. The accuracy with which a target can be located is determined by the beamwidth. We call such a pattern a pencil beam.

The circular aperture is a result of a paraboloid or of a lens antenna having circular symmetry, as in Fig. 5-15a and b. The field distribution in the cross section AA' arises from reflection or refraction of a section of energy between rays, as shown by the shaded area. In the section AA', the electric and magnetic fields may be considered to be in time phase. The amplitude in that section is determined from a radial diminution of energy from a point source (or other feed antenna). This is an approximate method, being quite good for reflectors and lenses which are large compared with wavelength.

Another approach for the reflector case consists in estimating the currents that flow in the reflector because of the electromagnetic wave being reflected there. In this case, at the reflector the tangent component of the electric field reduces to zero, whereas the tangent magnetic field becomes double that of the incident wave, as would be the case of a short-circuiting plane on a parallel-plane transmission line. The tangential magnetic field may be related to the surface-current flow in the reflector. From this current, the far-zone field can be calculated by using expressions of the type in (39) and (40).

EXERCISE

5-11 Derive the beamwidth expression for a circular aperture (67), and show that the first side-lobe power level is 17.6 db below that of the main beam.

5-8. REFLECTORS FOR A GIVEN POLAR DIAGRAM

For certain applications it is required to determine the aperture distribution for a given polar diagram. The problem is generally difficult if high gain is required as well. For reasonable gain requirements a superposition of a number of distributions of different amplitudes and appropriate phases suffices. Another approach is to shape the reflector so that with the primary feed the combined effect is to produce the desired pattern.

The patterns illustrated in Fig. 5-18 are called cosecant patterns since the radiation intensity is proportional to csc θ. This pattern is useful in radar detection of land targets by aircraft. If the aircraft is at height h, it is desired that for any angle θ, the field intensity received be independent of θ, the angle of elevation. Since the range R is equal to $h \csc \theta$ and since the electric field density is proportional to $1/R$ (each way), the polar diagram must have E proportional to $\csc \theta$ (or power proportional to $\csc^2 \theta$). Objects at 0 or 0' have approximately equal illumination.

In calculating far-zone fields, one encounters integrals of the form which can be expressed as Fourier integrals. It is often possible to express the field or current distribution as the Fourier transform of the far-zone field component. In other words, given a far-zone field, the aperture distribution (or current distribution) may be obtained. Unfortunately, since in addition to the amplitude the phase of the far-zone field must be known, the procedure is not practical. If one assumes the phase

Fig. 5-18 Cosecant radiation patterns. (*a*) Distributed sources; (*b*) shaped reflector.

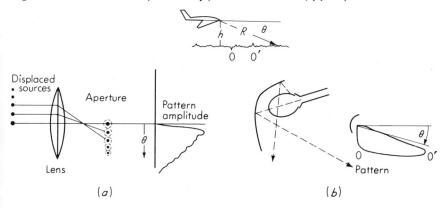

of the far-zone field, the aperture distribution may or may not come out to be physically realizable.

EXERCISE

5-12 An azimuth antenna for a GCA (ground control approach) system is shown.
a If $N = 150$ dipoles, estimate the width of the beam in azimuth and elevation degrees between 0.707 points.
b What should be the phase shift ψ_0 between dipoles in order to swing the beam 20° in the ϕ direction?

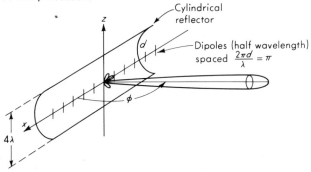

5-9. IMPEDANCE OF TRANSMITTING ANTENNAS

When radiating energy, a transmitting antenna is necessarily a dissipative load with respect to the generator and the network to which it is connected. We shall consider briefly the circuit aspects of the antenna and some of the factors which determine its efficiency. The antenna is a transducer between the network and space, but from the standpoint of the network the antenna appears to be a two-terminal load.

For our purposes we may think of the antennas as being connected to the rest of the system by a transmission line, such as a two-wire line or a coaxial line or a rectangular waveguide. Such is actually the case for the region of the spectrum above the medium-frequency region and is also the situation encountered in some systems at lower frequencies. We can restrict our attention then to the relation between the antenna and the transmission line, and ask about the progression of the system of waves from the transmission line through the antenna structure (or more properly stated, around the antenna structure) into space. We are interested in how the behavior of the system varies with frequency, that is, the bandwidth of the system.

As discussed in Chap. 2, a fundamental characteristic of waves is that they suffer reflection whenever there is a change in the structure of the system (that includes changes in media). The more abrupt the change, the larger, generally, is the amplitude of the reflected wave; generally speaking, whenever we encounter reflected waves, we encounter frequency sensitivity. We may, therefore, state one design principle as follows: Avoid discontinuity in structure as much as possible, and where changes in structure are necessary, the change should, if possible, be effected gradually over several wavelengths of path. Figure 5-19 shows several instances of how this principle can be applied.

As we go from the transmission line to the antenna, we pass through a region where the current system is very complicated in form, being neither wholly transmission-line currents nor wholly antenna currents. This is a region where there is usually a considerable amount of stored energy. As we go back into the line away from the antenna, the current system (or, if one prefers the language of field theory, the field distribution) becomes simpler in form and reduces to purely transmission-line distributions composed of a pair of waves, one going toward the antenna and the other coming from the antenna. The ratio of the amplitude of the reflected wave to the incident wave is called the reflection coefficient. The phase of the wave varies with position. Since the wave going in the $+x$ direction toward the antenna has a phase function

$$e^{-j\beta x} \qquad \beta = \frac{2\pi}{\lambda} = \frac{\omega}{v_p}$$

and the wave going in the $-x$ direction has a phase function

$$e^{+j\beta x}$$

the reflection coefficient varies with position as

$$\Gamma = \Gamma_0 e^{+2j\beta x} \qquad (68)$$

where Γ_0 is the reflection coefficient at the point taken to be $x = 0$. We are using β as the propagation constant, rather than the constant k of the radiation problem, because the propagation constant may be different on the transmission line from what it is in space. The most important

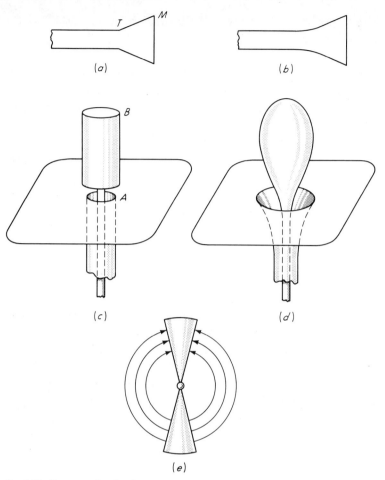

Fig. 5-19 Sources of reflections and methods of eliminating them. (a) Horn antenna with discontinuities at throat and mouth; (b) horn with throat discontinuity eliminated; (c) stub antenna over a ground plane with discontinuities at the transition from coaxial line to outside and at top of the stub; (d) broadband stub with both discontinuities smoothed out; (e) biconical antenna with point generator shows how antenna guides waves from generator and shows discontinuity experienced when waves reach ends of the cones and meet free space.

thing to note is how frequency-sensitive the reflection coefficient can be with respect to a change in its phase over a fixed interval $l = x_2 - x_1$.

The presence of a reflected wave in the transmission-line region is equivalent to an impedance mismatch or, shall we say, is representative of an impedance mismatch at the termination of the line. The relationship between the impedance (looking toward the load) and the reflection coefficient is

$$\Gamma = \frac{Z - Z_0}{Z + Z_0} = \frac{(Z/Z_0) - 1}{(Z/Z_0) + 1} \tag{69}$$

where Z_0 is the characteristic impedance of the line. The reflection coefficient can be extrapolated back to the neighborhood of the antenna and the equivalent impedance determined therefrom, but it must be remembered that the quantity is an extrapolated result and does not necessarily represent the "driving-point impedance" of the antenna. For example, where precisely would one set the driving point of the horn antenna of Fig. 5-19b? It is preferable to take a reference point in the transmission line, where the reflection coefficient can be determined cleanly, and regard that to be the location of the input terminals to the antenna. At that point we can speak of the input impedance as determined by (69) and separate the energy dissipated in all of the system beyond the terminals (reference point), covering both energy dissipated into heat by virtue of the fact that the conductors and dielectrics are not ideal ones and the energy radiated into space by the antenna.

The impedance given by (69) determines the power transfer from the generator in the usual manner. Referring to Fig. 5-20, we note that at the chosen reference point we see an impedance Z_A looking toward the antenna and an impedance Z_g looking toward the generator whose emf is V_g. Thévenin's theorem is applicable to this situation, and if V_g is the open-circuit voltage at the reference terminals O, the generator and line to the left can be replaced by a generator of emf V_g and internal impedance Z_g applied directly to the reference terminals. The condition for maximum power transfer is that

$$Z_A = Z_g^* \tag{70}$$

where the asterisk denotes the complex conjugate.

It should be understood that Z_g is the impedance seen at the refer-

Fig. 5-20 Equivalent circuit for an antenna and transmission line.

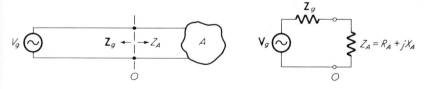

ence point O, not at the actual generator terminals. It is derived from the latter reference point by the transformation effected by a length of line. It may appear then that the relationship between the antenna and generator impedances with respect to power transfer would be dependent on the choice of reference point. However, as we move along the line, the impedance looking toward the antenna and the impedance looking toward the generator transform in a related way such that if the conjugate relationship (70) is satisfied at one point on the line, it is satisfied at all points (on the uniform line).

In order to minimize the effect of line length on the impedance, both the antenna and the generator are matched to the transmission line at the closest possible locations. Then if for some reason it is necessary to make a change in the length of line between the generator and the antenna, the conjugate impedance condition will remain satisfied. The matching networks are introduced as close as possible to the antenna or generator so that frequency sensitivity is minimized. From a wave point of view, the matching network produces a reflection in the line that cancels the reflection from, say, the antenna. Suppose this is accomplished at some one frequency. Now suppose the frequency changes. The reflection from the antenna which has to travel the distance between the antenna and the matching network undergoes a change in phase [see (68)] and will not then be canceled by the network reflection. The network reflection also changes in phase and amplitude with frequency, but the line-length effect is usually the dominant factor and the mismatch develops rapidly with frequency.

We have spoken about the antenna impedance in very general terms, but with respect to what one encounters directly in a practical situation. It would be desirable to be able to determine the impedance, given the antenna structure, from theoretical considerations. A very complete study has been made in the case of wire antennas. Schelkunoff has made a particularly engaging study of such antennas in terms of the biconical antenna (Fig. 5-19e) as a basic model. When the cone angles are very small, the biconical model approaches a thin wire antenna; when the cone angles are large, we develop from the model the spherical dipole antenna which is used in broadband systems (see Fig. 5-19d). Note that in the biconical model the generator is represented by a point generator at the apex of the cone. When the antenna is connected to a transmission line, new effects are added associated with which there is a storage of electric and magnetic energy.

As Fig. 5-19e indicates, the pair of cones form a waveguide for the waves from the generator. At the ends of the cones there is a discontinuity in the structure and the waves suffer reflection. The generator thus sees a mismatched transmission system. The discontinuity at the terminals has associated with it a storage of electric and magnetic energy; the storage of charges and the capacitance between the cones is immediately evident. Thus the system appears to the generator to have a resistance corresponding to the energy that passes on to the space around the antenna and the energy dissipated by the ohmic resistance of the cones as well as a reactance corresponding to the stored energy. The stored energy is determined by the geometry of the structure, and it is possible to have at some frequency the situation where the electric and magnetic energies are just equal. The difference between them vanishes, and the reactance is correspondingly equal to zero. The antenna is then resonant. Note that the theoretical resonant value will not correspond to the practical value, which is observed with respect to the transmission line, because of the added effects of the coupling region between the antenna and the line.

Another important result of the theoretical treatment of antennas is that the current distribution is found to be more and more closely represented by the sinusoidal distributions given in Fig. 5-6, the smaller the cone angles.

The reader is referred to Schelkunoff and Friis and to King for more (indeed, very much more) material on wire antennas. He will find there the complete story of how the input impedance of the idealized point-driven antenna depends on its length and transverse dimensions.

A concept that floats around in antenna engineering is the "radiation resistance." Consider first the dipole antenna. The power radiated is given by (26) in terms of the dipole moment. Thus

$$W = \frac{k^2}{12\pi} \left(\frac{\mu}{\epsilon}\right)^{\frac{1}{2}} |I_0|^2 l^2 \tag{71}$$

The same power would be dissipated by a resistance R_r carrying a current of the same amplitude; namely,

$$\frac{|I_0|^2 R_r}{2} = \frac{k^2}{12\pi} \left(\frac{\mu}{\epsilon}\right)^{\frac{1}{2}} |I_0|^2 l^2 \qquad R_r = \frac{k^2}{6\pi} \left(\frac{\mu}{\epsilon}\right)^{\frac{1}{2}} l^2 \tag{72}$$

This resistance is the radiation resistance.

In the case of the dipole antenna, where the current is distributed uniformly over the structure and is the current at the input terminals, the radiation resistance is the resistive component, except for ohmic losses in the dipole, of the impedance seen at the antenna terminals. The total impedance of the structure is incidentally highly reactive because it is associated with the capacitance between the loading structures at the ends of the antenna.

The radiation resistance is also used with reference to wire antennas such as those shown in Fig. 5-6. The current distributions are not uniform in the more general cases, and the definition is evidently somewhat artificial. The accepted procedure is to take the maximum value of the current along the wire as the reference current and determine the radiation resistance as that resistance which carrying the corresponding current would dissipate power equal to the power radiated by the antenna. Despite the peculiar nature of the radiation resistance, it serves as a sort of figure of merit that indicates whether other losses will be significant with respect to the radiation loss and thus is a measure of the efficiency of the antenna.

EXERCISE

5-13 Consider the half-wavelength antenna above a perfectly conducting earth. By examining the fields, particularly the electric field on the earth's surface, deduce what the radiation resistance must be if it is known that the resistance for the full half-wave antenna in empty space is 73 ohms. Also write a formula for E_θ at any point above the earth.

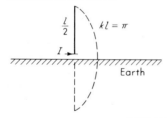

5-10. RECEIVING ANTENNAS

The discussion of receiving antennas is greatly simplified by a beautiful theorem that relates the receiving characteristics to the radiating characteristics. Figure 5-21a is a representation of a receiving system. The antenna, which is taken to be the whole system to the

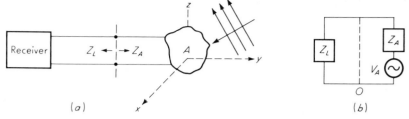

Fig. 5-21 Receiving antenna equivalent circuit.

right of the reference terminals O, has a transmitting impedance Z_A as discussed in the previous section. The antenna now feeds into the line and finally to a load (representing all the rest of the receiver) whose impedance, as it appears at the terminals O, is Z_L. A plane wave falls on the antenna from some direction θ, ϕ.

First of all, we must remember that the antenna radiates a definite *polarization* in the direction of θ, ϕ which means that there is a corresponding orientation of polarization of the incident wave to which the antenna will have the maximum response. In the case when the antenna radiates linear polarization, the best response is obtained when the incident wave is linearly polarized and the polarization coincides with that on transmission. If the polarization of the incident wave makes an angle ψ with the polarization on transmission in the given direction, the amplitude of the incident wave that acts on the antenna is reduced by the factor $\cos \psi$.

The incident wave on the antenna excites a system of currents and thus excites a wave in the transmission line. From the standpoint of the antenna itself, it is again a transducer between space and the line. From the standpoint of the line and the load, the antenna is a generator feeding the system. In fact the theory shows that if we take the open-circuit voltage V_A developed at O under the incident wave, the power relationships for the system are those corresponding to a generator of emf V_A and internal impedance Z_A applied at the terminals O.

Using the Poynting theorem and reciprocity considerations, we can show that the open-circuit voltage induced at the terminals of a receiving antenna is as follows:

$$V_{20} = -\frac{1}{I_2(0)} \int_{\substack{\text{free}\\\text{space}}} \mathbf{J}_2 \cdot \mathbf{E}_{21}\, dV$$

The electric field $\mathbf{E}_{21}$ is produced by a distance source at the antenna under consideration. $\mathbf{J}_2$ is the current density distribution on the antenna, and $I_2(0)$ is the total current at the driving point. For a simple linear wire antenna, the current distribution is obtained from Eq. (41), and the induced voltage expression becomes

$$V_{20} = -\frac{1}{I_2(0)} \int_{-l/2}^{+l/2} I(z) E_{z_{21}}\, dz$$

It is obvious that for an infinitesimal dipole antenna the open-circuit voltage is $V_{20} = -E_{21} l \cos \psi$, where the angle ψ is the tilt angle of the antenna with respect to the electric field.

The condition for maximum power transfer, the antenna and incident wave being fixed in this connection, is just the usual relationship between a generator and the load, as follows:

$$Z_L = Z_A^* \tag{73}$$

The power delivered to any load is obtained by applying the equivalent circuit of Fig. 5-21b.

The power delivered to the load is the power absorbed from the wave incident on the antenna from outside. The absorption of power can be characterized by an absorption cross section, as though the antenna presented a certain area to the oncoming wave and absorbed all the power from the wave falling on that area. If the power absorbed is W_{abs} and the power per unit area of the incident wave is P, the absorption, or receiving, cross section of the system (antenna and given load) is

$$A_R = \frac{W_{\text{abs}}}{P} \tag{74}$$

All other things being fixed, the cross section is a function of the load and the simple computation of how the power varies with Z_L on the basis of the equivalent circuit of Fig. 5-21b will yield the dependence of the cross section on the load.

For a fixed system (antenna and load) the response varies with the direction of incidence of the wave. The absorption, or receiving, cross section is thus a function of aspect. The functional dependence on aspect, for optimum orientation of the incident polarization for each aspect, is known as the receiving pattern of the system. The all-important theorem is that the receiving pattern of the system is the same as the transmitting pattern of the system. Thus we arrive at the concept of the cross section as a function of θ,ϕ. Correspondingly, we can talk about the average value of the cross section over all possible aspects, that is, the average of the function $A_R(\theta,\phi)$ over all solid angles. Hence

$$\langle A \rangle = \frac{1}{4\pi} \iint A_R(\theta,\phi) \sin\theta \, d\theta \, d\phi \tag{75}$$

The average cross section also depends on the impedance relationship between the antenna and the load. Another remarkable theorem on radiating systems is that the average cross section for systems that are matched, that is, for which the antenna and load satisfy (73), is a universal constant

$$\langle A_0 \rangle = \frac{\lambda^2}{4\pi} \tag{76}$$

Finally, we see without too much difficulty that, as a result of the reciprocity theorem concerning radiation and transmitting patterns, the cross section presented by a matched system in the direction θ,ϕ is related to the value of the transmitting gain function of the receiving antenna in that direction by

$$A_R{}^0(\theta,\phi) = g(\theta,\phi) \frac{\lambda^2}{4\pi} \tag{77}$$

The result as given can be used without further ado for any linearly polarized antenna. When the antenna is elliptically polarized (as it may be in some particular directions, at least), the above result cannot be applied without great care as to the meaning of the gain. We shall have to leave the matter at this stage and refer the reader to treatises on antennas for further treatment of generalized antennas.

EXERCISES

5-14 What is the open-circuit voltage of a half-wavelength antenna oriented in a field of strength 5 μv/m to produce maximum voltage? What is the maximum possible power that the antenna can deliver to the load at a frequency of 100 Mc? What is it at a frequency of 10 Gc?

5-15 Calculate the effective area for a half-wave dipole antenna. Compare this result with the effective area for the infinitesimal dipole.

5-16 Two dipole antennas are oriented in space with their equivalent circuits as shown. By calculating the area A of the receiver according to the load power absorbed, verify directly that $g/A = 4\pi/\lambda^2$. $R_r = 20(kl)^2$ and $k = 2\pi/\lambda$.

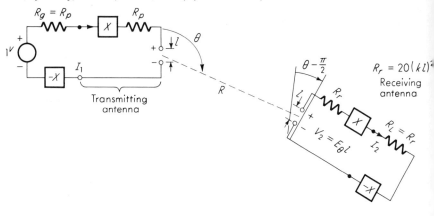

5-11. SPECIAL REFLECTOR-TYPE ANTENNAS USED IN MICROWAVE COMMUNICATION SYSTEMS

In many involved communication systems the equipment becomes quite complicated and massive. Because of this and because long transmission lines are generally bothersome as to attenuation and phase changes, it is often desirable to locate the receiver or at least the front end of the receiver as close to the antenna feed as possible. If the antenna is used with low-noise receivers such as parametric amplifiers or masers, the equipment is inserted at the horn feed. This can be done easily in several ways.

The *Cassegrain antenna*[6] used for pencil-beam production consists of a main dish which is a paraboloid and an auxiliary reflector or subdish which is a hyperboloid (see Fig. 5-22a). In addition to the advantage stated above, it is possible to obtain an overall axial dimension which is less than that of the paraboloid of equivalent aperture but with no subdish. The feed in this case would be at the virtual feed position. A problem which does arise, however, is that of aperture blocking due to the presence of the subdish. If necessary, this may be overcome by making the subdish a wire grating so oriented that the wave emerging from the horn is reflected into the main dish. At the main dish, a polar-

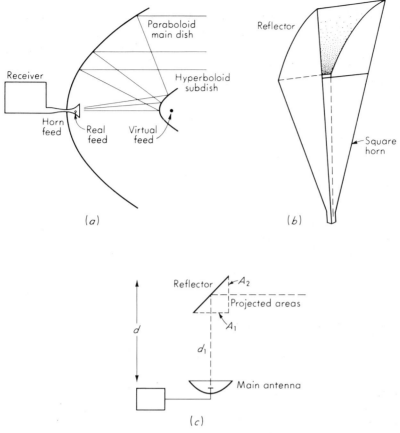

Fig. 5-22 Special reflector-type antennas. (a) Cassegrain antenna; (b) horn reflector (hog-horn) antenna; (c) periscope antenna.

ization twister reflects the wave oriented so that its polarization is orthogonal to that of the horn wave. This polarization permits the passage of the wave through the subdish and reduces the blocking effect.

In order to minimize the problems due to far-side and back-lobe interference, a combination horn and reflector antenna has been devised called a *horn-reflector* or *hoghorn antenna*.[11] The reflector is a portion of a paraboloid and is made an integral part of the antenna placed in front of a square horn (see Fig. 5-22b). For use in space communications, the antenna is characterized by the shielding of the horn, which prevents noise power from being received from the ground.

Certain microwave relay systems make use of passive reflectors placed in the distant field of the transmitting and receiving antenna to redirect the incident electromagnetic wave. Figure 5-22c shows an application of this principle as used in a periscope antenna. This arrangement may be used to clear some obstruction with the total length of the feeding waveguides kept to a minimum. In addition to the loss due to the length of the path from one antenna to the next, which would normally be there if the antennas were facing each other, there is that loss due to the finite projected areas A_1 and A_2. If the projected areas of the reflector are as large as the main beam, this loss may be quite small. The system has also the disadvantage of possible interference with other transmission paths. The illuminated reflector, being finite, acts as an excited antenna structure radiating in directions in addition to the specular one. Making the reflector a portion of a paraboloid whose focus is at the main antenna (because of its focusing action) increases the gain of the periscope antenna by a few decibels.

EXERCISES

5-17 Calculate power received corresponding to a transmitted power of 100 watts and a distance between transmitter and receiver of 10^3 m under the following conditions:

a Gain of transmitting antenna = 1.5; effective area of receiver = 0.40 m².
b Gain of both antennas = 2; wavelength = 0.10 m.
c Effective area of both antennas = 1 m²; wavelength = 0.03 m.

5-18 For the data of Exercise 5-17(**b**) and the radiation resistance of both antennas equal to 50 ohms, calculate $|Z_{12}|$. Now, allowing separation r to vary, find the separation for which Z_{12} becomes comparable to Z_{11}.

5-12. FREQUENCY-INDEPENDENT ANTENNAS

As we have seen in this chapter, the properties of commonly used antennas can be very sensitive to changes in frequency. An antenna which has characteristic frequency-independent properties is a biconical antenna. Unfortunately, the fact that the antenna must be terminated raises some problems. A class of antennas which has frequency-independent[12] properties can be obtained by specifying the antenna shape by angle only as follows: Let the surface of the antenna be specified by

$$r = F(\Phi) \tag{78}$$

where r is the radial distance and Φ is an angle measured from a reference axis. It must be pointed out that the shape of the antenna can be three-dimensional, in which case an additional angle θ would be needed for specifying the shape, as follows:

$$r = F(\theta, \Phi) \tag{79}$$

To illustrate this type of antenna, only a planar structure is presented below. If we scale to a new frequency which is $1/K$ times the original frequency, a point on the antenna must lie at a radial distance corresponding to

$$r' = KF(\Phi) \tag{80}$$

where K is not a function of the angle Φ. When we effect this change, we need the surfaces to be congruent by a rotation in the angle Φ. Then

$$KF(\Phi) = F(\Phi + C) \tag{81}$$

where C is the angle through which the scaled antenna must be rotated to become congruent to the original antenna. C depends on K but not on Φ. To obtain the functional relationship for the shape of the antenna, differentiate (81) with respect to C. Thus

$$\frac{\partial K F(\Phi)}{\partial C} = \frac{\partial F(\Phi + C)}{\partial C} = \frac{\partial F(\Phi + C)}{\partial (\Phi + C)} \tag{82}$$

and now differentiating (82) with respect to Φ yields

$$\frac{K \, \partial F(\Phi)}{\partial \Phi} = \frac{\partial F(\Phi + C)}{\partial \Phi} = \frac{\partial F(\Phi + C)}{\partial (\Phi + C)} \tag{83}$$

Combining (83) and (82) yields

$$\frac{\partial K}{\partial C} F(\Phi) = K \frac{\partial F(\Phi)}{\partial \Phi} \tag{84}$$

$$\frac{1}{K} \frac{\partial K}{\partial C} = \frac{1}{r} \frac{\partial r}{\partial \Phi} = a \tag{85}$$

from which we obtain

$$r = F(\Phi) = r_0 e^{a(\Phi - \Phi_0)} \tag{86}$$

which is the equation of an equiangular spiral. Figure 5-23 illustrates the shape of such an antenna surface. Examining (86), we may illustrate the frequency independence as follows: Let r_1 indicate a radial distance at a wavelength λ_1, such that

$$r_1 = r_0 e^{a(\Phi_1 - \Phi_0)} \tag{87}$$

Then at a new wavelength λ_2 this point should be at a position

$$r_2 = \frac{\lambda_2}{\lambda_1} r_1 \tag{88}$$

Substituting (88) into (86), we obtain

$$r_2 = \frac{\lambda_2}{\lambda_1} r_0 e^{a(\Phi_1 - \Phi_0)} = r_0 e^{\ln(\lambda_2/\lambda_1)} e^{a(\Phi_1 - \Phi_0)} = r_0 e^{a[\Phi_1 - \Phi_0 + (1/a)\ln(\lambda_2/\lambda_1)]} \\
= r_0 e^{a(\Phi_2 - \Phi_0)} \tag{89}$$

The final term of (89) shows that the effect of a change in λ is equivalent to a rotation. It can be shown with a bit of algebra that taking a cut

Fig. 5-23 Equiangular spiral antenna.

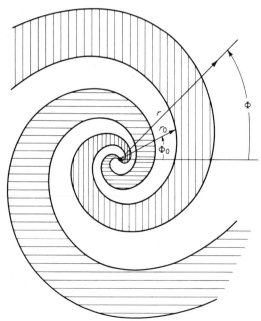

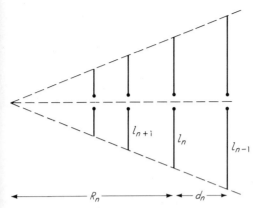

Fig. 5-24 Basic log-periodic antenna structure.

through the antenna structure with Φ equal to a constant gives

$$\frac{r_2}{r_1} = \frac{e^{a(\Phi_1-\Phi_0+2\pi m)}}{e^{a[\Phi_1-\Phi_0+2\pi(m-n)]}} = e^{2\pi na} \qquad (90)$$

This structure is called log periodic since it is periodic in the exponent. Figures 5-24 and 5-25 are sketches of log-periodic antenna structures. We see that

$$\frac{l_n}{l_{n-1}} = \frac{d_n}{d_{n-1}} = \tau$$

Typical values of τ range from 0.8 to 0.9. It is possible to combine a frequency-sensitive array with a log-periodic array to produce a fre-

Fig. 5-25 Log-periodic antenna structure.

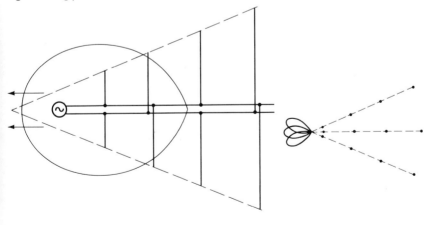

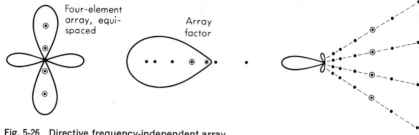

Fig. 5-26 Directive frequency-independent array.

quency-independent, highly directive antenna array. An example[13] of such a combination is shown in Figs. 5-26 and 5-27.

REFERENCES

1 SILVER, S. (ed.): "Microwave Antenna Theory and Design," MIT Radiation Laboratory Series, vol. 12, McGraw-Hill Book Company, New York, 1949.

Fig. 5-27 Feed system for array.

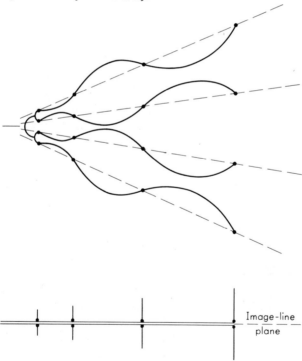

2 KRAUS, J. D.: "Antennas," McGraw-Hill Book Company, New York, 1950.

3 FRY, D. W., and F. K. GOWARD: "Aerials for Centimetre Wave-lengths," Cambridge University Press, London, 1950.

4 HULL, G. F., JR.: Application of Principles of Physical Optics to Design of U.H.F. Paraboloid Antennas, *Bell Lab. Rept.* MM-43-110-2, Feb. 8, 1943.

5 RONCHI, L., and G. TORALDO DI FRANCIA: An Application of Parageometrical Optics to the Design of a Microwave Mirror, *IRE Trans.*, vol. AP-6, pp. 129–133, January, 1958.

6 WOODWARD, B.: The Cassegrain Antenna, *Proc. IRE,* vol. 46, p. 2A, March, 1958.

7 MARTINDALE, J. P. A.: Lens Aerials at Centimetric Wavelengths, *J. Brit. IRE,* vol. 13, pp. 243–259, May, 1953.

8 SMITH, R. A.: "Aerials for Metre and Decimetre Wave-lengths," Cambridge University Press, London, 1949.

9 VON AULOCK, W. H.: Properties of Phased Arrays, *Proc. IRE,* vol. 48, pp. 1715–1727, October, 1960.

10 STEGEN, R. J.: Excitation Coefficients and Beamwidths of Tschebyscheff Arrays, *Proc. IRE,* vol. 41, pp. 1671–1674, November, 1953.

11 CRAWFORD, A. B., et al.: A Horn-reflector Antenna for Space Communications, *Bell System Tech. J.,* vol. 40, no. 4, p. 1095, July, 1961.

12 RUMSEY, V. H.: Frequency Independent Antennas, *IRE Natl. Conv. Record,* Part 1, pp. 114–118, 1957.

13 MEI, K. K.: A Broadside Log-periodic Antenna, *Proc. IEEE,* vol. 54, no. 6, pp. 889–890, June, 1966.

6

propagation of radio waves

6-1. INTRODUCTION

Except for the brief remarks in the introduction to Chap. 1, we have said nothing about the effect of environmental factors on the performance of antennas and on communication between systems. We shall consider the subject of propagation briefly and try to call attention to the most significant factors.

The propagation of electromagnetic waves around the earth is influenced by the properties of the earth and its atmosphere. The earth is an inhomogeneous body whose electromagnetic properties vary considerably as we go from one point to another. Sea water is highly conducting, whereas the desert sands are dielectric, having virtually zero conductivity but dissipating energy by virtue of polarization losses. The atmosphere over the earth is a dynamic medium, its properties varying with temperature and humidity; turbulence creates blobs that scatter radiation. In the upper atmosphere we have regions of a high degree of ionization and large numbers of free electrons. The ionosphere behaves like a highly conducting medium over a large range of frequencies, and in that range, waves striking the ionosphere are reflected and return to the earth's surface. The ionosphere is influenced by the sun and indeed owes its existence to the sun, and consequently its properties undergo diurnal variations and seasonal variation. Activity of the sun soon becomes evident by the change in the ionosphere and the corresponding change in propagation and communication. The ionosphere is also a turbulent medium, and anomalous scattering phenomena are the result. It should also be noted that the earth's magnetic field plays a significant role in the behavior of the ionosphere. The various factors that we have mentioned make long-range communication possible on the one hand and are factors resulting in interference effects and fading on the other. The phenomena that are observed are rather strongly frequency-dependent and are determined also by the type of antenna that is used.

Figure 6-1 (Ref. 2) shows schematically how the different parts of the environment come into play in the problem of wave propagation. The figure considers the case when there is a line-of-sight path from the transmitter T to the receiver R. In that case, as is shown, there is a direct wave from the transmitter to the receiver determined by the free-space pattern of the antenna. Another route is by way of a reflection from the ground or, more properly stated, from the earth's surface, whatever it happens to be. The amplitude of the reflected wave is determined by the properties of the earth in the region of interest. The

Fig. 6-1   Radio-wave propagation.

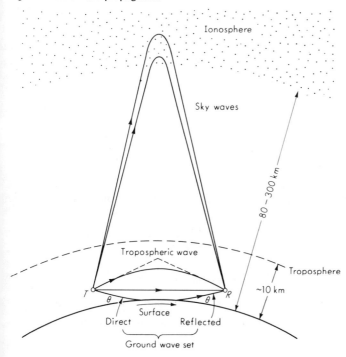

propagation of radio waves

density and composition of the troposphere vary with height, and (as an associated phenomenon) the index of refraction varies. In an inhomogeneous medium the rays are not straight lines. Also, the troposphere can show some stratification under certain conditions, which also results in bending of the rays. Finally, turbulence can occur with the attendant phenomenon of scattering. All of the factors mentioned can contribute to energy reaching the receiver by a tropospheric path. Another contribution to the field at the receiver can come by way of the ionosphere. As indicated by the diagram, the reflection by the ionosphere is not really sharp. The situation is complicated further by the fact that the ionosphere has a layer structure. However, when it is all said, the result is that over certain frequencies it is possible to get a sky-wave contribution to the field. All of the contributions must be added with due regard for their relative amplitudes and phases. The amplitudes are determined by the directivity of the antenna, by the reflection characteristics of the various regions, and by the distance attenuation factor. The relative phases are determined also by the reflection characteristics in addition to the relative path lengths traversed from the transmitter to the receiver. The contributions may interact constructively to produce a field intensity that is greater than what would exist under free-space conditions, or they may interact so as to cancel one another and produce a lower field intensity (in fact, virtually zero field intensity). Since the factors involved vary with the time of day and the weather, the result is a fluctuating field intensity.

6-2. THE GROUND-WAVE SET

As is shown in Fig. 6-1, the ground-wave set may be said to be composed of a space-wave group (direct and reflected waves) and a surface wave. This decomposition is done so that the highly complex state of affairs resulting from the analysis of K. A. Norton[3] may be made more useful.

Following Bullington's procedure,[4] the field due to the ground-wave set is expressed as

$$\frac{E}{E_0} = 1 \quad + Re^{j\Delta} \quad + (1-R)Ae^{j\Delta} + \cdots \tag{1}$$

| Direct wave | Reflected wave | Surface wave | Induction field and secondary effects of the ground |

The reflection coefficient R of the ground depends on the angle of incidence and polarization of the wave and on the ground characteristics. It is discussed more fully later on in the section. The angle Δ is the phase difference resulting from the difference in length of the reflected and direct rays. A study of Fig. 6-2 shows that

$$\Delta = \frac{2\pi}{\lambda}(L_2 - L_1) = \frac{2\pi}{\lambda}\left[\left(\frac{h_T + h_R}{d}\right)^2 + 1\right]^{\frac{1}{2}} - \frac{2\pi d}{\lambda}\left[\left(\frac{h_R - h_T}{d}\right)^2 + 1\right]^{\frac{1}{2}} \quad (2)$$

For a distance d between antennas greater than about five times the sum of the two antenna heights h_T and h_R, the angle

$$\Delta \approx 4\pi \frac{h_T h_R}{\lambda d} \quad (3)$$

E_0 is the peak free-space field intensity at a distance d from an antenna of gain g_T transmitting a power of W_T watts. From Chap. 5, Eqs. (27) to (33), it can easily be seen that

$$g_T = \frac{d^2 4\pi}{W_T} \frac{E_0^2}{120\pi} \frac{1}{2} \quad (4)$$

from which

$$E_0 = \frac{(60 W_T g_T)^{\frac{1}{2}}}{d}$$

This expression, for a short vertical dipole antenna with transmitted power $W_T = 1$ kw and gain $g = 1.5$, becomes $e_0 = 300$ mv/m at 1 km from the source.

The surface-wave attenuation factor A in (1) depends on the polari-

Fig. 6-2 Direct and reflected wave.

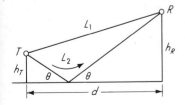

165
propagation of radio waves

zation and ground constants. The phrase *surface wave* is here used as defined in Refs. 3 and 4. It should not be confused with the Sommerfeld or Zenneck surface wave.[2]

A plot of the curve for the attenuation factor A in terms of a numerical distance p and a phase constant depending on the earth's characteristics is given in Fig. 6-3 (p. 627 of Ref. 2). For $\theta \approx 0$, the following approximations may be made:

$$b \approx \tan^{-1} \frac{K_r + 1}{K_i} \quad \text{and} \quad p \approx \frac{\pi d}{\lambda K_i} \cos b \tag{5}$$

where $\epsilon = \epsilon_0(K_r - jK_i)$ represents the complex permittivity for which $K_i = \sigma/\omega\epsilon_0 = 18 \times 10^3 \sigma/f_{\text{Mc}}$ and K_r is the relative dielectric constant. In K_i and hence in p, σ is the conductivity of the earth, λ is the wavelength, and f_{Mc} is the frequency (in megacycles per second) of the wave.

For $p < 1$, $A \approx 1$ and $E_0 \approx e_0/d = 300$ mv/d, whereas for $p > 10$, A is proportional to $1/d$ and E is proportional to e_0/d^2. It can easily be shown that

$$p \propto \frac{f^2 d}{\sigma} \tag{5a}$$

for low frequency or good conductive earth and that

$$p \propto \frac{fd}{K_r + 1} \tag{5b}$$

Fig. 6-3 Surface-wave attenuation factor.

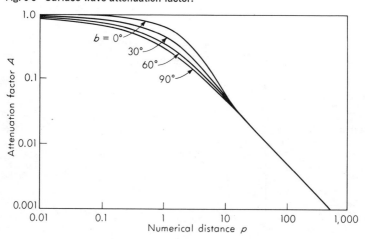

for high frequency or high-dielectric-constant earth. The earth's curvature can be neglected for surface-wave consideration up to a distance of $50/f_{\text{Mc}}^{\frac{1}{3}}$ miles.

Since the wave penetration into the earth (skin depth) ranges from about 100 ft at broadcast frequencies down to a few feet at VHF, rain and other surface phenomena have little effect on the surface-wave propagation.

The Space Wave

The direct wave and the ground-reflected wave shown in Fig. 6-1 constitute the space-wave components of the ground-wave set. It is evident that they are meaningful only when there exists a line of sight between the transmitter and receiver but lose significance when the receiver is below the horizon of the transmitter.

The ground-reflected wave follows the optical law of reflection that the angle of incidence (angle between incident ray and the normal to the surface) equals the angle of reflection. Thus the ground-reflected wave appears to come from an image antenna. The antenna over the earth is not a simple image problem, however, because the earth is not a perfect conductor and because the surface of the earth is curved. Correction for the latter factor is a relatively simple matter, but the effect of finite conductivity is more complicated. The amplitude and phase of reflection vary from point to point, and consequently the image is effectively a function of the reflection point.

The properties of a lossy medium, when the losses are associated with conductivity and with electric polarization, are characterized by a complex permittivity

$$\epsilon = \epsilon_r - j\epsilon_i \tag{6}$$

The permittivity is made up of dielectric (polarization) effects and conductivity. The conductivity contribution is

$$\epsilon_i = \frac{\sigma}{\omega} \tag{7}$$

where σ = conductivity and ω = angular frequency. Alternatively, the medium is characterized by ϵ_r and the loss tangent

$$\tan \delta = \frac{\epsilon_i}{\epsilon_r} \tag{8}$$

The reflection characteristics are given in terms of plane-wave reflection coefficients for a flat earth. There are two fundamental cases of polarization of the wave, vertical (in the plane of incidence) and horizontal (normal to the plane of incidence). In each case let θ be the grazing angle, the angle between the incident ray and the ground. Then the reflection coefficient (pp. 140–141 of Ref. 2) is

$$R = \frac{E_{\text{ref}}}{E_{\text{inc}}} \tag{9}$$

The ratio of the complex amplitude (magnitude and phase) of the reflected wave to that of the incident wave is

1. Vertical polarization

$$R_v = \frac{(K_r - jK_i)\sin\theta - [(K_r - jK_i)\cos^2\theta]^{\frac{1}{2}}}{(K_r - jK_i)\sin\theta + [(K_r - jK_i)\cos^2\theta]^{\frac{1}{2}}} \tag{10}$$

2. Horizontal polarization

$$R_h = \frac{\sin\theta - [(K_r - jK_i) - \cos^2\theta]^{\frac{1}{2}}}{\sin\theta + [(K_r - jK_i) - \cos^2\theta]^{\frac{1}{2}}} \tag{11}$$

where

$$K_r = \frac{\epsilon_r}{\epsilon_0} \qquad K_i = \frac{\epsilon_i}{\epsilon_0} \tag{12}$$

ϵ_0 being the permittivity of free space.

The complex reflection coefficients give the shift in phase as well as the change in amplitude in reflection. When the grazing angle θ is small ($\theta \approx 0$),

$$R_v \approx -1 \qquad R_h \approx -1 \tag{13}$$

that is, the reflection takes place with a phase shift of 180° and no change in amplitude. For angles of incidence other than zero, when the conductivity gets very large so that $K_i \gg K_r$ and $K_i \gg 1$,

$$R_v \approx +1 \qquad R_h \approx -1 \tag{14}$$

This is the case over seawater, for example, for frequencies up to the order of 1,000 Mc. (See Fig. 6-4.)

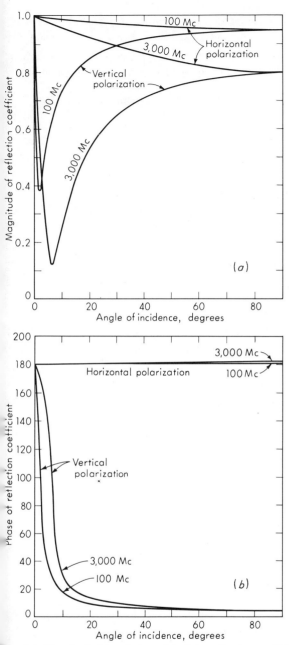

Fig. 6-4 Magnitude (a) and phase (b) of the reflection coefficient as a function of angle of incidence (for seawater). The phase of the reflected wave lags the phase of the incident wave. (*After Burrows and Atwood.*)

When the transmitter and receiver antennas are near the ground, the limiting values (13) can be used for separations d that are large compared with both h_T and h_R, the heights of the transmitter and receiver, respectively, above the earth. The field strength at the receiver then varies as

$$E_f = \frac{2e_0}{d} \sin \frac{2\pi h_T h_R}{\lambda d} \tag{15}$$

$$E_f \approx \frac{M}{d^2} \quad \text{for} \quad \frac{2\pi h_T h_R}{\lambda d} \ll 1 \tag{16}$$

where $M = e_0(4\pi h_T h_R/\lambda)$.

Thus the field attenuates more rapidly than it would from an antenna in free space. The derivation of (15) assumes that the patterns of the antennas are broad so that there is negligible difference between the gain in the direct-ray direction and the reflected-wave direction. For highly directive antennas the assumption is no longer valid. It should be realized that in the microwave region beams are now being obtained having half-power widths less than 1°.

Consider a transmitter-receiver at a height h_2 and a target at height h_1 over a good reflector and at a distance d away. The power at the target is

$$W_1 = \frac{ae_0^2}{d^2} 4 \sin^2 \left(\frac{2\pi}{\lambda} \frac{h_1 h_2}{d} \right)$$

where a is a constant. The power scattered by the target and sent to the receiver at height h_2 is

$$W_2 = K_s W_1 \frac{4 \sin^2}{d^2} \frac{2\pi h_1 h_2}{\lambda d}$$

where K_s is a scattering constant.

Now in the absence of the reflecting surface

$$W_{1_0} = \frac{ae_0^2}{d^2} \quad \text{and} \quad W_{2_0} = K_s W_{1_0} \frac{1}{d^2}$$

The ratio of power received via the space wave, including reflection, to the power received over the direct wave is

$$\rho = \frac{W_2}{W_{2_0}} = 16 \sin^4 \frac{2\pi h_1 h_2}{\lambda d} \tag{17}$$

which for small values of $2\pi h_1 h_2/\lambda d$ reduces to

$$\rho \approx 16 \left(\frac{2\pi h_1 h_2}{\lambda d}\right)^4 \tag{17a}$$

An inspection of the equation for fixed heights and distance indicates the advantage of using short-wavelength radio waves.

EXERCISES

6-1 Two aircraft 2.0 km apart are flying low over the ocean, their height above water being 10 m.
a What is the field strength at the second aircraft if the first radiates 250 watts of power and has an antenna of gain 1.5 over a half-wave radiator? The frequency is 300 Mc. Show that the surface wave is negligible.
b Same as (a), but the aircraft are both at 15 km above water.

6-2 A transmitting antenna and a receiving antenna are within sight of each other, and the intervening ground is to be considered perfectly flat.
a What is the objection to having the antennas very close to the ground?
b What is the maximum line-of-sight range for a transmitting antenna raised 100 m above the ground? Ignore refractive-index variations.

6-3 An antenna transmits 150 watts and has a gain (power) $g(\theta) = 5 \cos \theta$. The antenna is horizontally polarized and is situated 1,000 m from another antenna (receiver).
a In the absence of any reflections, what is the field intensity at the receiver position? The effective receiving area of the receiving antenna is $A_0 = 0.5$ m². What is

$f = 3 \times 10^6 c$

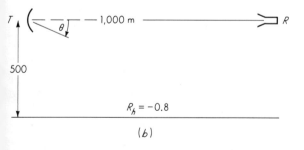

the power received in the matched load? What would be the power received if the antenna load is not matched?

b Now consider that both antennas are 500 m above a smooth surface for which the reflection coefficient is $R_h = -0.8$. What is the field intensity at the receiver position?

6-4 An antenna is located above a "perfectly conducting ground." A signal is received at a point A some distance away from the antenna.

a Would it be possible for the received signal to have the *same* value as that arriving in the absence of the ground? Explain.

b Would it be possible for the received signal to have a larger value than that arriving in the absence of the ground? Explain.

6-3. TROPOSPHERIC WAVE

The effect of the atmosphere can be resolved into the three phenomena of refraction, reflection, and scattering. *Reflection* takes place when the atmosphere becomes stratified. This is actually a not infrequent meteorological phenomenon, and when it occurs communication is possible over distances considerably greater than the line of sight.

The normal condition of the atmosphere, if there is anything really normal about meteorological conditions, is a temperature and pressure gradient with which is associated a gradient of index of refraction and consequently a *refracted* wave. The varying index of refraction results in a bending of the rays, and the sense of the gradient is such that rays are bent toward the earth. Again we can obtain communication beyond the line of sight. Changes in temperature and moisture content result in changes in the gradient of the index of refraction. Conditions may result in an increase in index with height and a bending of the rays away from the earth. The signal intensity at a station where reception depends significantly on tropospheric waves then fades markedly. Conversely, meteorological conditions can result in a steeper gradient such that rays are bent toward the earth more strongly. This condition arises when a body of warm air is over water, resulting in a high density of water vapor over the surface. Under such conditions the atmosphere forms a "duct," or waveguide, that guides the waves over the surface. Abnormally large ranges beyond the line of sight are then obtained. The effect is significant in the microwave region.

To illustrate this refraction phenomenon, let us assume an atmosphere for which the index of refraction decreases linearly with height.

Thus

$$n = n_\alpha(1 - \beta y) \tag{18}$$

The index of refraction is defined as $n = c/v$, where $c \approx 1/(\mu_0\epsilon_0)^{1/2}$ is the velocity of the wave in vacuum and $v = 1/(\mu\epsilon)^{1/2}$ is the velocity in a medium of permeability μ and permittivity ϵ. In (18), n_α is the value of n at $y = 0$, and β is a proportionality constant. (See Fig. 6-5a.)

A ray starts at an angle α with respect to the horizon at each earth level and makes an angle θ at height y. Snell's law gives

$$n_\alpha \cos \alpha = n_\alpha(1 - \beta y) \cos \theta$$
$$1 - \beta y = \sec \theta \cos \alpha$$
$$-\beta \, dy = \cos \alpha \sec \theta \tan \theta \, d\theta$$
$$-\beta \frac{\cos^2 \theta}{\sin \theta} \sec \alpha = \frac{d\theta}{dy} \tag{19}$$

However, from the geometry of Fig. 6-5b, $1/R = K$ is the curvature and $R \, d\theta \sin \theta = dy$, and combined with (19) this gives

$$K = -\beta \cos^2 \theta \sec \alpha \quad \text{(curvature)} \tag{20}$$

For small angles of incidence, θ and α are small and

$$K \approx -\beta$$

However, from (18)

$$\frac{dn}{dy} = -n_\alpha \beta$$

Fig. 6-5 Refraction of waves. (a) Decrease in density; (b) ray geometry.

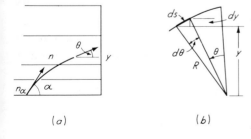

(a)   (b)

propagation of radio waves

from which $K = -\beta = (1/n_a)(dn/dy)$, and since $n = K_r^{1/2}$, where K_r is the relative dielectric constant,

$$\frac{dn}{dy} = \frac{1}{2K_r^{1/2}} \frac{dK_r}{dy}$$

and finally

$$K = \frac{1}{n_a} \frac{dn}{dy} \approx \frac{1}{2K_r} \frac{dK_r}{dy} \tag{21}$$

which is half the lapse rate of the permittivity.

Turbulence in the atmosphere results in inhomogeneities of index of refraction and *scattering*[8] of the waves. The scattering becomes especially pronounced when the wavelength of the electromagnetic wave is of the same order of magnitude as the characteristic turbulence length (i.e., the dimensions of the atmospheric blobs). This scattering phenomenon is not to be confused with scattering by clouds or rain, which is also significant in certain parts of the spectrum. Scattering causes both fading and enhanced reception depending on the wavelength, positions of the transmitter and receiver, and antenna sizes. Following the procedure of Booker and Gordon[7] as given in an early paper, we may interpret scatter propagation as the scattering of electromagnetic waves by a turbulent medium consisting of inhomogeneities in the atmosphere. An expression for scattered power per unit solid angle, per unit incident power density, and per unit macroscopic element of volume is derived using an autocorrelation point of view. Although the analysis is beyond the scope of this text, it is instructive to proceed a bit further. Denote the scattering factor by $\sigma(\Theta,X)$. The angles are defined so that the scattering in a particular direction makes an angle Θ with the direction of incidence and an angle X with the direction of the incident electric field. The factor $\sigma(\Theta,X)$ is shown to be a function of $\overline{(\Delta K_r/K_r)^2}$, the mean-square fractional deviation of dielectric constant from the average, a function of the mean wavelength in the medium, and a function of the scale of the turbulence. The scattering factor has, therefore, a directional effect depending on all the parameters mentioned.

Consider now the transmission over a distance by means of beamed antennas as shown in Fig. 6-6.

The two beams intersect in a common volume of turbulence.

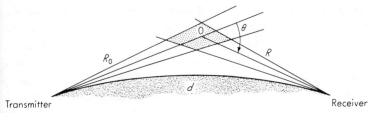

Fig. 6-6 Scatter propagation geometry.

Assume that both the transmitting and receiving antennas have the same dimensions and area A. Assuming that the antennas are large compared with wavelength, the gain is, from (62) (Chap. 5),

$$g = \frac{4\pi A}{\lambda^2}$$

Now consider the scattering of a macroscopic element of volume dV at the point Q. If the transmitted power is W_T, the power scattered by dV is

$$dW_s = \frac{W_T}{4\pi R_0^2} \frac{4\pi A}{\lambda^2} \sigma(\Theta, X) \, dV \qquad (22)$$

and the corresponding total power at the receiver is

$$W_R = \int \frac{W_T}{4\pi R_0^2} \frac{4\pi A}{\lambda^2} \sigma(\Theta, X) \frac{A}{R^2} \, dV \qquad (23)$$

where the integration is with respect to dV over the volume of atmosphere common to both transmitting and receiving beams. This integration is difficult because of the dependence of σ on Θ and X and because of the shape of the common volume.

If the received power W_R is measured in picowatts (10^{-12} watt), the transmitted power W_T in kilowatts, the distance d between transmitter and receiver in kilometers, the area A of the antennas in square meters, the wavelength λ in meters, and the modified scale of turbulence l' in meters, (23) becomes

$$W_R = W_T \frac{4A}{\lambda d} \left(\frac{2\pi l'}{\lambda}\right)^3 \qquad (24)$$

and this formula gives the scattered power received between nearly identical antennas and for many practical cases of beam communication. It should be remembered that under certain conditions of wavelength, distance, dielectric constant, etc., the power may be scattered in such a direction as to fall outside the receiving beam. For typical conditions, the signal received by scattering will be combined with that received in the absence of turbulence in the atmosphere. The resultant signal will undergo considerable fading, depending on the amplitudes and phases of the direct and scattered components.

EXERCISE

6-5a Prove that the radius of curvature of an electromagnetic "ray" traveling in the troposphere is given by $R = -(dn/dh)^{-1}$, where n is the refractive index at height h. (Note that this equation is true only for rays whose directions are close to being parallel to the earth's surface.)
b If the diameter of the earth is 12,740 km, what variation of refractive index would give rise to a ray traveling entirely parallel to the earth's surface?

6-4. SKY-WAVE PROPAGATION

For frequencies which are sufficiently low, the ionosphere offers still another means of increasing the range of communication. In order to show the relation between the ionization and the frequency at which reflections may occur, ionospheric transmission is considered as follows.

The radiation from the sun ionizes the upper atmosphere. The degree of ionization is a function of pressure, and the resulting effect of the competing processes of absorption of the ionizing radiation and recombination is that the ionization is intense over the band of altitude 80 to 300 km and falls off in both directions from that band. Below 50 km the ionization is insignificant as far as radio-wave propagation is concerned.

The ionization shows a certain degree of stratification, and we speak of D, E, F_1, and F_2 layers in increasing order of height, and as a consequence the layers move up and down during the course of day and night. Figure 6-7 shows an approximate distribution of the densities.

Effective Dielectric Constant of the Ionized Region

We assume that there are no collisions between the molecules and the electrons in the medium and that we can neglect the effect of the

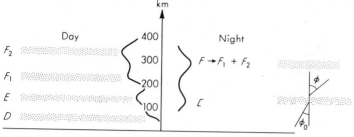

Fig. 6-7 A layer model of the ionosphere.

earth's magnetic field. Then it is possible to represent the presence of the electrons as if the electromagnetic wave propagated through an equivalent dielectric region.

The effective dielectric constant is determined as follows:

f = m**a**

$eE = m \dfrac{d\mathbf{v}}{dt}$

If $E = E_m \cos \omega t$, $eE_m \cos \omega t = m \, dv/dt$, from which the velocity of the electron is obtained as

$$v = \dfrac{e}{\omega m} E_m \sin \omega t$$

The convection-current density is

$$i_c = Nev = \dfrac{Ne^2}{\omega m} \sin \omega t$$

and the displacement-current density is

$$i_d = \dfrac{dD}{dt} = \epsilon_0 \dfrac{dE_m}{dt} \cos \omega t = -\epsilon_0 \omega E_m \sin \omega t$$

These combine to form a total current density of

$$i_t = \left(\dfrac{Ne^2 E_m}{m \omega} - \epsilon_0 \omega E_m \right) \sin \omega t$$

From this, an equivalent displacement-current density is defined as

$$i = -\epsilon \omega E_m \sin \omega t \quad \text{where } \epsilon = \epsilon_0 \left(1 - \frac{\omega_p^2}{\omega^2}\right)$$

The plasma radian frequency depends on the density of electrons and is $\omega_p = (Ne^2/m\epsilon_0)^{\frac{1}{2}}$. Upon substitution of the electron constants and ϵ_0, the equivalent dielectric constant becomes

$$\epsilon_r = 1 - \frac{81N}{f^2} \tag{25}$$

If N is expressed in terms of the number of electrons per cubic centimeter, f is in kilocycles per second. The index of refraction $n = \epsilon_r^{\frac{1}{2}}$ approaches unity for $f^2 \gg 81N$, which is generally obtainable at *microwave* frequencies. For $f^2 < 81N$, the index of refraction becomes imaginary.

Reflection from the Ionosphere

As in the case for tropospheric waves, Snell's law applies. Thus

$$n \sin \phi = n_0 \sin \phi_0$$

where ϕ is measured from the vertical at a point at which the electron density is such as to yield an index of refraction of n. At the lowest region of the ionosphere, n reduces to n_0, which is unity since the electron density is taken as zero at the entrance to the ionosphere. Bending of the ray so as to produce reflection occurs at $\sin \phi = 1$, for which case the ray path is parallel to the ionospheric layer. For a vertical ray ($\phi_0 = 0$), $n = 0$, and $(1 - 81N/f^2)^{\frac{1}{2}} = 0$; that is, $f_v^2 = 81N$. We shall call the critical frequency the minimum vertical ray frequency for which no reflection occurs in a given layer because we have reached a maximum electron density (Figs. 6-8 and 6-9) of

$$f_c^2 = 81N_{\max} \tag{26}$$

If $f \leq f_c$ for all frequencies less than the critical frequency, reflections occur for all angles ϕ_0 since $\sin \phi_0$ is imaginary. If $f > f_c$, then what can be said? Solving for f from above, we obtain $f = 9N^{\frac{1}{2}} \sec \phi_0 = f_c \sec \phi_0$.

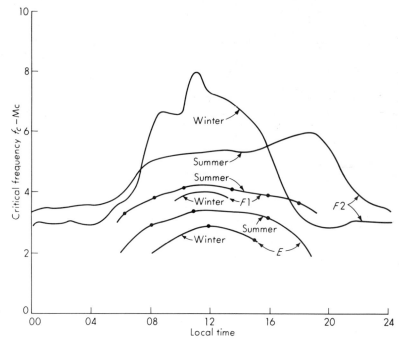

Fig. 6-8 Typical curves of the diurnal variation of critical frequency for the various ionosphere layers.

Hence, if f is less than $9N^{\frac{1}{2}} \sec \phi_0$, we can get reflections. This frequency $9N_{\max} \sec \phi_0$ is called the maximum usable frequency (MUF). The maximum usable frequency is the greatest frequency of a wave that is bent back by the layer. It depends on the angle of the ray and is a minimum for a vertical ray. Because of the simple relation between the critical frequency and the maximum density of ionization ($f_c = 9N_{\max}^{\frac{1}{2}}$) and because it is more convenient to measure the critical frequency directly, the critical frequency is generally quoted instead of the maximum density.

The vertical height (virtual height) of a layer is given by the height A above the earth at which a perfect reflector can be imagined as reflecting the ray arriving in the same manner as it does from the ionosphere. The time for a ray to travel the path from the transmitter to the "reflector" and thence to the receiver is the same time as a ray going through the ionosphere experiences. The shorter curved path is traveled at a lesser group velocity. Measuring the time that a pulse takes to travel the vertical height measures the virtual height. From the geometry (Fig.

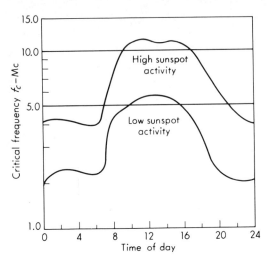

Fig. 6-9 Typical variation of critical frequency with solar activity.

6-10), it is seen that

$$\sec \phi_0 = \frac{1}{h'/[h'^2 + (d/2)^2]^{\frac{1}{2}}} = \frac{f}{f_c} \tag{27}$$

This relates the various parameters and allows for either the determination of the maximum usable frequency for a given distance or vice versa.

The concept of critical frequency is based on continuous variation of electron density with position. However, the ionosphere, like the troposphere, undergoes fluctuation in structure, a sort of turbulence that gives rise to blobs of ionization density. Such blobs scatter the radiation, and again enhanced transmission is possible over special spectral regions.

Fig. 6-10 Determination of virtual height.

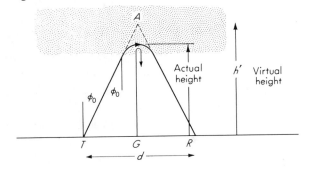

Some of the phenomena have been associated with meteors which produce local trails of intense ionization. The study of the scattering phenomenon has led to recognition of the conditions under which enhanced transmission is possible, and by effectively using high-gain antennas, long-range transmission has been effected in the UHF region.

EXERCISES

6-6 Assume that the ionosphere is a uniform layer lying between 100 and 300 km above the earth.

a If the electron density is $2.085 \times 10^{12}/m^3$, what frequencies will be reflected at vertical incidence?

b Radar measurements have been made in the region of 39 Mc in order to determine the distance from the earth to the moon and to some planets. Estimate the error introduced into this determination by the presence of an ionosphere of the idealized type specified in (a). Assume that the measurements are made when the moon or planets are exactly overhead. Recall that the radar pulses travel at the group velocity $v_g = c\epsilon_r^{1/2}$, where $c \approx 3 \times 10^8$ m/sec.

6-7 It is planned to establish communication between two points T and R which are 500 miles apart. Use is to be made of a satellite passing 500 miles above, as shown in the figure. Assume that the satellite scatters sufficient energy for all frequencies under consideration. The antennas are steerable (track the satellite) and have radiation on beams confined to a beam angle of $\pm 10°$. Over the total frequency

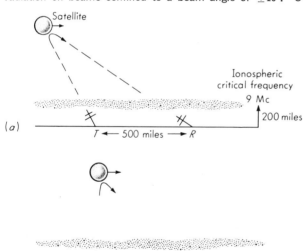

propagation of radio waves

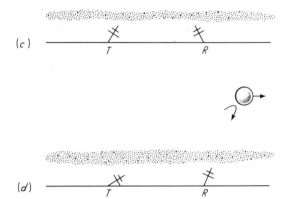

intervals given in **(a)** and **(b)**, trace the possible signal paths from the transmitter at T to the receiver at R as the satellite follows the path indicated in the figure.

a As a first attempt, existing ionosphere testing equipment (antennas, receivers, transmitters, etc.) of frequency range 2 to 30 Mc is to be used.

b The equipment is changed so as to operate in the frequency range of 1 to 5 Gc.

6-8 A ham radio enthusiast wishes to talk to Hawaii several evenings a week. He asks you, as a consulting engineer, to

a Decide what frequency he should use.

b Design the antennas at both locations.

c Decide what transmitter power is necessary.

d Explain your choices so he, an amateur, can understand (and believe) them; in particular, explain how propagation influences the choice of frequency and antenna.

e Tell him at what time in San Francisco he can hope to get the best reception to and from Hawaii.

6-5. RADIO-WAVE ABSORPTION IN ATMOSPHERIC GASES

In choosing the microwave frequency at which a communication is to operate, we must take care to allow for the attenuation of the energy in an atmosphere in the absence of precipitation as well as in the presence of rainfall, fog, and clouds.

Resonant absorption by neutral oxygen and uncondensed water vapor molecules is a salient characteristic of the atmosphere at centimeter and millimeter wavelengths. Neutral oxygen molecules are present in the atmosphere about 21 percent by volume and remain at about this concentration well into the stratosphere. On the other hand, H_2O vapor concentration of about 1 percent decreases (on the average) very rapidly with height.

The O_2 molecule is paramagnetic, and its permanent magnetic dipole moment interacts with the incident field at wavelengths of about 5 mm and at 2.5 mm. The H_2O vapor molecule interacts through its permanent electric dipole at 13.5 mm and 1.63 mm and at several other wavelengths of less than 1 mm.

Figure 6-11 shows the absorption (decibels per kilometer) for an atmosphere containing oxygen and water vapor at approximately sea level.

Precipitation affects microwave propagation in a complex way. Although fundamental absorption and scattering coefficients can be obtained simply in terms of the wavelength, drop size, and dielectric constant of water, a typical storm will be characterized by considerable variation in rainfall concentration and temperature as well as drop size. If the drop size is assumed to be less than 0.1 to 0.2 mm, the absorption is dependent primarily on wavelength and total mass of liquid water per unit volume in the air and is inversely proportional (roughly) to the square of the wavelength. For a temperature of 18°C, the absorption coefficient is approximately 0.005 (db/km)/(g/cm^3) at λ = 10 cm, increasing to 0.4 at 1 cm. For a typical 1 g/m^3 water content, the attenuation at 10 cm would be about 0.005 db for a 1-km cloud path.

REFERENCES

1 TERMAN, F. E.: "Electronic and Radio Engineering," 4th ed., chap. 22 (note slight difference in definition), McGraw-Hill Book Company, New York, 1956.

2 JORDAN, E. C.: "Electromagnetic Waves and Radiating Systems," chaps. 16–17, Prentice-Hall, Inc., Englewood Cliffs, N.J., 1950.

3 NORTON, K. A.: The Propagation of Radio Waves over the Surface of the Earth and in the Upper Atmosphere, Part I, *Proc. IRE,* vol. 24, p. 1367, October, 1936; Part II, *Proc. IRE,* vol. 25, pp. 1203–1236, September, 1937; The Calculation of Ground-wave Field Intensities over a Finitely Conducting Spherical Earth, *Proc. IRE,* vol. 29, pp. 623–629, December, 1941.

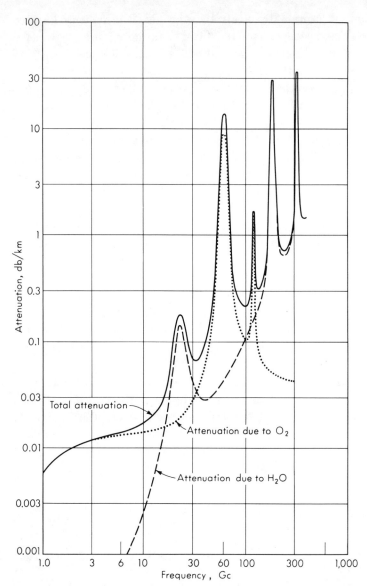

Fig. 6-11 Calculated atmospheric absorption at sea level.

4 BULLINGTON, K.: Radio Propagation at Frequencies above 30 Megacycles, *Proc. IRE,* vol. 35, p. 1122, October, 1947.
5 BREMMER, H.: "Terrestrial Radio Waves," Elsevier Publishing Company, Amsterdam, 1949.
6 PEDERSON, P. O.: "The Propagation of Radio Waves," G.E.C. Gad, Copenhagen, 1929.
7 BOOKER, H. G., and W. E. GORDON: A Theory of Radio Scattering in the Troposphere, *Proc. IRE,* vol. 38, pp. 401–417, April, 1950.
8 *Proc. IRE,* vol. 43, October, 1955, scatter propagation issue.
9 *Proc. IRE,* vol. 45, June, 1957, VLF propagation issue.
10 *Proc. IRE,* vol. 47, February, 1959, ionosphere, IGY issue.
11 BACHYNSKI, M. P.: Microwave Propagation over Rough Surfaces, *R C A Rev.,* vol. 20, pp. 308–335, June, 1959.
12 BURROWS, C. R., and S. S. ATTWOOD: "Radio Wave Propagation," Academic Press Inc., New York, 1949.
13 REED, H. R., and C. M. RUSSELL: "Ultra High Frequency Propagation," John Wiley & Sons, Inc., New York, 1953.
14 BEAN, B. R., G. D. THAYER, and B. A. CAHOON: Methods of Predicting the Atmospheric Bending of Radio Rays, *J. Res. Natl. Bur. Std.,* vol. 64D, pp. 487–492, September–October, 1960.
15 MILLMAN, G. H.: Atmospheric Effects on VHF and UHF Propagation, *Proc. IRE,* vol. 46, pp. 1492–1501, August, 1958.
16 VAN VLECK, J. H.: The Absorption of Microwaves by Oxygen, *Phys. Rev.,* vol. 71, pp. 413–424, April 1, 1947.
17 VAN VLECK, J. H.: The Absorption of Microwaves by Uncondensed Water Vapor, *Phys. Rev.,* vol. 71, pp. 425–433, April 1, 1947.

7

noise

7-1. INTRODUCTION

"Noise" in a communication system denotes undesired electrical excitations of various origins which reduce the information carried by the signal. These undesired excitations may be divided into two classes, those which are periodic in time and those which are random in time. The first class is characterized by a discrete frequency spectrum, often with only one or two main components, and thus is subject to elimination by filtering. The second class is characterized by a continuous spectrum, which often is not a function of frequency (at least in the frequency band of interest). The term *white noise* is used to denote random excitations which have a uniform frequency spectrum. Some random excitations have a spectral distribution such that the noise power is inversely proportional to frequency, and appropriately are called $1/f$ noise.

If the system designer is clever, periodic disturbances can always be eliminated (by filtering or by cancellation schemes, for example); the random excitations, however, cannot be entirely eliminated and thus impose a lower limit to the signal power which can be detected. This lower detectable limit of signal power depends on the accuracy desired in the communication system; it is a function of the system of modulation which is employed, the path of propagation between transmitter and receiver, the components used at both the transmitting and receiving stations, etc. Thus the design of the entire system must be based on the minimization of the noise which interferes with the transmitted information at the output of the receiver. One method of doubling the signal-to-noise ratio at the receiver is to double the transmitter power. Another method is to decrease the receiver noise by a factor of 2. If the transmitter power is in the megawatt range, an expensive receiver may be the more economical solution (by orders of magnitude!).

The purpose of this chapter is a physical and partially mathematical description of some of the more important kinds of random noise, for it is only through understanding noise that its effects may be minimized. Important parameters which describe the noise performance of an amplifier or receiver are discussed along with typical performance data for important microwave devices. Before proceeding to more detailed considerations, however, it is worthwhile to briefly list some important sources of noise and their characteristics.

Electrical motors, home appliances, induction heaters, automobile ignition, etc., are all *manmade sources of noise*. The undesired electrical disturbances from these sources may appear as line-voltage fluctuations, electromagnetic induction fields, or even radiation. The first two sources are often controlled by manufacturers, and the FCC regulates the amount of radiation allowed. The interference from the above sources is generally at specific frequencies and therefore can be eliminated by filtering.

Radio-frequency interference (RFI) is another source of manmade noise which is becoming more important with time, both in systems which use radiated electromagnetic waves and also in closed systems, where, for example, unwanted harmonics of one signal may interfere with the information processing of a second signal. RFI is generally not noticeable in a local urban area such as New York or San Francisco, but at a location midway between two transmitters at the same frequency, it is very bothersome.

Naturally occurring *electrical discharges*, such as lightning, produce noise. Many lightning flashes along a propagation path generate a uniformly spread-out noise, whereas one sharp flash near the receiver is more properly treated as a sharp impulse interference. Electrical discharges from airplanes produce noise similar to many lightning flashes. This noise is called precipitation static.

Sky noise has several sources. The sun is a strong source of noise but is very localized. Galactic noise is of more importance since it originates in all portions of the sky (although some portions are stronger than others). In addition, molecular absorption lines (due to oxygen and water vapor, primarily) are important at frequencies above 10 Gc. Noise power from the sun decreases inversely as the frequency, and galactic noise power decreases inversely as frequency squared. An important consequence of these dependences is a region between 2 and 8 Gc where an antenna pointed toward the sky receives less noise power than it would at either higher or lower frequencies. This "window" is especially important for satellite and space communications. Figure 7-1 illustrates noise from space expressed in terms of antenna temperature (see Sec. 7-6). The relation of sky noise to the system is considered in Sec. 7-8.

Thermal, or *Johnson* (after its discoverer), noise is caused by the thermal motions of electrons. If this motion occurs in a resistor, the sum of all electron velocities toward the left end of the resistor is a fluctuating quantity, depending upon the temperature, the collisions with atoms in the conductor, and the electric field which would exist if

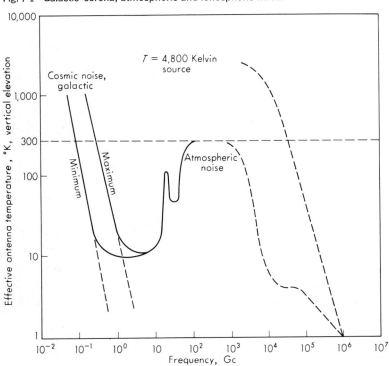

Fig. 7-1 Galactic corona, atmospheric and ionospheric noise.

microwave communications

more electrons were at one end of the resistor than the other. As a result of this last observation, at thermal equilibrium, the time average of the sum of all electron velocities is zero (in the absence of any applied voltage). However, the rms voltage across the resistor is not zero and consequently the resistor can deliver power as a result of the thermal motion of its electrons. This noise power is one of the most important sources of noise, because it is so fundamental. It is discussed in more detail in a later section.

Shot noise arises from the statistical fluctuation in the number of electrons making up the current flow in a given device, such as a vacuum tube or transistor. It also is very fundamental to communication systems and is discussed more extensively later.

Flicker noise in tubes is one form of $1/f$ noise; another form is found in transistors. This noise, sometimes called *excess noise*, is not well understood, and as it is important generally only at very low frequencies, it will not be discussed further here.

7-2. CHARACTERIZATION OF NOISE SOURCES

A study of noise is a study of a large number of events (time functions) which take place during a given time. If every event is independent of every other event, the noise is purely random and has zero correlation. When dealing with larger and larger numbers of independent events, it rapidly becomes impossible to deal with each individual event and then sum over all events. It is much easier and more accurate (as the number of events becomes very large) to deal with the probabilities describing a given event or, in other words, to use statistics. A study of noise, therefore, must be based on a thorough understanding of statistics. Since such a study of statistics is beyond the scope of this work and since many readers will not have a thorough grounding in statistics, the statistical results which we need will be introduced at the appropriate time.

One important property of randomly varying quantities which results from statistical studies† may be stated as follows: If two independent quantities having zero correlation are added together, the mean value of the resultant quantity is the sum of the two mean values. In a linear system, noise voltages have a mean or average value of zero; thus when

† See, for example, A. Van der Ziel, "Noise," Prentice-Hall, Inc., Englewood Cliffs, N.J., 1954.

two noise voltages are superimposed, their mean-square values are added or, stated slightly differently, their *powers* add (just as do the powers of two orthogonal voltages at the same frequency or the power of two voltages at different frequencies). This property is basic to the treatment of electrical noise.

It proves convenient to characterize a noise source by an equivalent circuit, either a current source shunted by an admittance or a voltage source in series with an impedance. Once this characterization is properly performed, we can then use circuit analysis to determine the noise voltage, current, or power at a particular point in the circuit, just as we do for a signal. It is well to stress, however, that the equivalent current source or voltage source is specified by a mean-square current or voltage and that the calculations which we will perform will yield only a mean-square current or voltage at the point of interest. This suggests that noise power is a more fundamental quantity than equivalent noise voltages and currents, which could rapidly lead into a philosophical argument. We merely state that in microwave applications, where sources are matched to transmission lines, which are in turn matched to amplifiers, antennas, and the other components of the system, noise power is a much more useful parameter than noise currents and voltages. Conversely, in vacuum-tube amplifiers at lower frequencies, where maximum voltage transfer rather than maximum power transfer is desired, the equivalent noise-voltage characterization of a noise source is more useful. The two characterizations are really equivalent, and one should be readily derivable from the other (except possibly in special limiting cases, as discussed below).

Consider a voltage generator in series with an impedance Z, as shown in Fig. 7-2a. The current generator shunted by an admittance Y, as shown in Fig. 7-2b, can be used equally well to characterize the noise source, if the impedance Z is finite. If the real part of Z is $R(f)$ and the real part of Y is $G(f)$, the available power from these two representa-

Fig. 7-2 Equivalent noise sources.

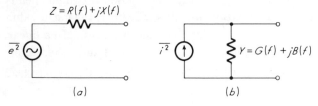

tions of the same noise source in a bandwidth df may be written as

$$w(f)\,df = \frac{d\overline{(e^2)}}{4R} = \frac{d\overline{(i^2)}}{4G} \tag{1}$$

which defines $w(f)$, the mean available power per unit bandwidth at frequency f. Both the voltage generator and the current generator in Fig. 7-2 generate random time functions. It should be emphasized that, by describing $w(f)$ as the mean *available* power per unit bandwidth, we assume the circuit representing the noise source is matched to its load and therefore is delivering maximum power to the load. Equation (1) can be used to define the mean-square noise current in terms of the mean available power per unit bandwidth. Thus

$$d\overline{(i^2)} = 4G(f)w(f)\,df \tag{2}$$

where the frequency dependences have been written explicitly. Note that, for the case of a noise-current source, when $G \to 0$, $w(f)$ must become large without limit. In this limiting case, the concept of available power is not useful, and instead, the actual power delivered to a finite load will be considered. The mean-square noise voltage may be similarly defined.

In a microwave system, the frequency bandwidth which determines the noise present in the signal is often set by the intermediate-frequency (IF) amplifier, which is typically at 30 Mc. As these amplifiers are generally voltage or current amplifiers, their transfer characteristic is specified in terms of a *transmittance*, which is the complex ratio of an output variable (voltage or current) to an input variable (voltage or current). The transmittance may be the transfer impedance, transfer admittance, transfer-voltage ratio (voltage gain or loss), or transfer-current ratio (current gain or loss).[1] In general, the transmittance is a function of frequency. For noise calculations, an equivalent, or "noise," bandwidth is often defined in terms of an idealized filter transmittance, which has a constant amplitude in a given frequency range and zero amplitude outside this range, as shown in Fig. 7-3a. Figure 7-3b shows a typical filter transmittance, and superimposed on it (dashed lines) is the equivalent idealized noise transmittance. The equivalent idealized noise transmittance has the same maximum value as the filter transmittance and delivers the same mean-square total output voltage or

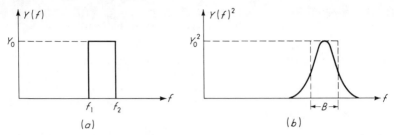

Fig. 7-3 Equivalent filter transmittance. (a) Idealized filter transmittance; (b) noise bandwidth B from filter transmittance.

current from a white-noise source as the filter transmittance. This result can straightforwardly be justified by circuit theory and means mathematically that the noise bandwidth B is

$$B = \frac{1}{Y_0^2} \int_0^\infty |Y(f)|^2\, df \tag{3}$$

To provide some insight into the meaning of (3), consider the circuit of Fig. 7-4. The transmittance $Y(f)$ of the box is defined by the equation

$$\overline{e_o} = Y(f)\overline{i_i} \tag{4}$$

where $\overline{i_i}$ is a sinusoidal signal input current at frequency f. To find the mean-square output voltage, multiply (4) by its complex conjugate, which yields

$$\overline{e_o e_o^*} = Y(f)Y(f)^* \overline{i_i i_i^*} \tag{5}$$

Suppose that $\overline{i_i}$ is the input current from a resistance, i.e., a noise current. As such, only the mean-square value has a finite time average, and therefore we should expect only a mean-square value for the output voltage. In a later section we shall show that for most frequencies a resistance at temperature T is characterized by a constant mean

Fig. 7-4 Transmittance $Y(f)$.

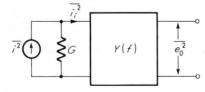

available power per unit bandwidth, or $w(f) = kT$, where k is Boltzmann's constant and T is the absolute temperature. Assuming the resistance is independent of frequency (i.e., the parasitic capacitances have negligible effect at the frequency of interest), from (2) we have

$$\overline{di^2} = 4GkT\,df$$

and, by substituting this into (5), we obtain for an incremental frequency range df (when the conductance G matches the circuit) the following:

$$\overline{de_o^2} = GkT\,Y(f)^2\,df \tag{6}$$

This may be integrated immediately over all frequencies to obtain the mean-square output voltage

$$\overline{e_o^2} = GkT Y_0^2 \left[\frac{1}{Y_0^2} \int_0^\infty |Y(f)|^2\,df \right] \tag{7}$$

In this last equation, the term before the brackets represents the maximum mean-square output voltage in any given incremental bandwidth, and the quantity in brackets is the noise bandwidth.

It is convenient to define the noise bandwidth in a slightly different manner for microwave systems for which the input and output of components such as amplifiers, attenuators, etc., are generally matched to a transmission line. An easily measured parameter of such a component is the power transfer function, or power gain. If W_i denotes the input power at a given frequency and W_o the output power at that frequency when the amplifier is matched at both input and output, the available power gain is given by

$$W_o = A(f)W_i \tag{8}$$

where $A(f)$ is a positive, real number. The thermal-noise output power in such a case, in an incremental bandwidth df, is

$$dN_o = kTA(f)\,df \tag{9}$$

and the total noise output power may then be written as

$$N_o = A_o kT \left[\frac{1}{A_o} \int_0^\infty A(f)\,df \right] \tag{10}$$

where the quantity preceding the brackets denotes the maximum noise output power and the quantity in brackets denotes the noise bandwidth.

7-3. THERMAL NOISE

Electrical fluctuations due to the thermal motion of electrons in conducting materials generally set the lower bound on the detectable signal voltage, current, or power in a communication system, and therefore can be called the most fundamental limitation in the amplification of weak electrical signals. Any derivation of the available power or equivalent mean-square thermal voltage of a resistor is based on the Maxwell-Boltzmann equipartition law as modified by Planck. To gain any understanding of thermal noise, therefore, we should thoroughly understand this equipartition law. A mathematical derivation of this law would take us far afield, so we shall merely state what the law says and briefly attempt to understand what it means. The interested reader is referred to one of the standard works of statistical mechanics for a derivation.

Consider a large number of monatomic gas molecules in a container. These molecules will bounce off the walls and each other, and in the steady state the number in the box will remain constant (on the average, none are absorbed by the walls of the box). The energy of one atom with velocity v and mass m is

$$U = \tfrac{1}{2}m(v_x^2 + v_y^2 + v_z^2) \tag{11}$$

The energy of any other molecule may be similarly expressed. The Maxwell-Boltzmann equipartition law requires that the energy of a system be represented by the summation of a large number of terms N, each proportional to the square of a different generalized coordinate necessary to describe the total configuration of the system. The law states that if such a system is in thermal equilibrium, the total energy of the system when averaged over a long period of time is $\tfrac{1}{2}kTN$. Here k is Boltzmann's constant, 1.37×10^{-23} joule/°K, and T is the absolute temperature. At a standard temperature of 290°K,

$$kT = 4 \times 10^{-21} \text{ joule}$$

a convenient number to remember.

Each generalized coordinate which enters into the energy of the system quadratically is called a *degree of freedom*. Thus the atom mentioned above has three degrees of freedom, corresponding to the x, y, and z components of velocity, respectively. A one-dimensional harmonic oscillator, whose energy is given by

$$U = \tfrac{1}{2}mv_x^2 + \tfrac{1}{2}kx^2 \qquad (12)$$

has two degrees of freedom, corresponding to one velocity component and one position coordinate. Thus a generalized coordinate can be either a velocity or a position, so far. The energy of an LC circuit can be written as

$$U = \tfrac{1}{2}Li^2 + \tfrac{1}{2}Cv^2 \qquad (13)$$

and here the generalized coordinates are current and voltage, respectively. Therefore, the energy per degree of freedom, according to the Maxwell-Boltzmann equipartition theorem, is $\tfrac{1}{2}kT$. Applications of the Maxwell-Boltzmann equipartition theorem gave results in agreement with theory in most cases; for example, the theorem correctly predicted the blackbody radiation law at lower frequencies but fell down at the higher frequencies. In 1900, Planck resolved this discrepancy (and laid an important foundation store for modern physics) by postulating that energy was quantized in "bits" of hf; he showed that, under this assumption, the energy per degree of freedom should be modified by a factor $p(f)$ such that

$$p(f) = \frac{hf/kT}{e^{hf/kT} - 1} \qquad (14)$$

where h = Planck's constant = 6.62×10^{-34} joule-sec. The factor $p(f)$ does not deviate appreciably from unity, except for very high frequencies. For example, at 600 Gc and with $T = 290°$, $p(f) = 0.95$.

The energy per degree of freedom is now $\tfrac{1}{2}kTp(f)$.

We are now ready to derive the available thermal-noise power from a resistor; because of our emphasis on microwaves, we shall follow a proof set down by Nyquist originally, but we could equally well use a circuit-theory approach. Consider a lossless transmission line of length L, matched at each end by lossy terminations (see Fig. 7-5). These

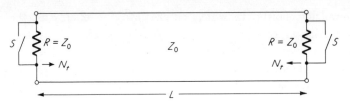

Fig. 7-5 Terminated transmission line.

terminations may be envisaged as resistors R which equal the characteristic impedance Z_0 of the transmission line. Both the transmission line and the terminations are at temperature T. Thermal-noise power N_t from the left-hand resistor will flow down the line toward the right-hand resistor and vice versa. Since the system is in thermal equilibrium, these two powers will be equal. This power flows as electromagnetic waves which are generated by fluctuating currents in the respective terminations. The power from the left-hand termination in a small frequency band Δf will be called ΔN_t.

Suppose both ends of the transmission line are simultaneously shorted by closing the switches S. The power flowing in the line at the moment the switches are closed is trapped. The shorted line becomes a resonant cavity, with many possible resonant modes; we shall consider only the transverse electromagnetic (TEM) modes. The TEM resonances occur when the line is an integral number of half wavelengths long, or at frequencies

$$f_n = \frac{c}{\lambda_n} = c\,\frac{n}{2L} \qquad (15)$$

The trapped noise power becomes stored energy which propagates back and forth along the line in the allowed resonant modes. The stored energy in a small bandwidth Δf which is taken large enough to include several resonant modes is

$$\Delta U = 2\,\frac{\Delta N_t L}{c} \qquad (16)$$

where L/c is the time required for a TEM wave to travel the length of the line; the factor 2 is required because thermal power was traveling in both directions when the line was shorted.

We now may use the equipartition theorem to determine ΔU and thus find ΔN_t, which is our objective. The energy stored on the transmission line is stored in the electric and magnetic fields, which represent two degrees of freedom for each mode. The number of modes in a bandwidth Δf is determined from (15) as

$$\Delta n \leq \frac{2L}{c} \Delta f \tag{17}$$

where Δn is an integer. If the mode spectrum is plotted, as in Fig. 7-6, the reason for the less-than-or-equal-to sign in (17) becomes apparent. The stored energy in a frequency band Δf at thermal equilibrium is given by the equipartition theorem as

$$\Delta U = 2[\tfrac{1}{2}kTp(f)] \Delta n \leq kTp(f) \frac{2L}{c} \Delta f \tag{18}$$

If we let the length of the line L be arbitrarily long, Δn will be a large integer even for an arbitrarily small frequency interval Δf.

In this case we can eliminate the less-than signs from the above equations, and by using (18), we obtain

$$dN_t = kTp(f)\, df \tag{19}$$

where differentials have replaced the Δ's. Thus the mean available power per unit bandwidth for thermal noise is

$$w(f) = kTp(f) \tag{20}$$

The available power in an incremental frequency band df from a matched resistor at temperature T in thermal equilibrium with the circuit to which it is connected is given by (20). For low frequencies,

Fig. 7-6 Energy spectrum due to noise on shorted transmission line.

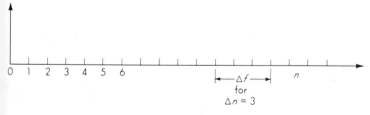

$hf/kT \ll 1$ and $p(f) \approx 1$, and this reduces to the usual formula for thermal noise, $kT\,df$. At higher frequencies, $p(f)$ decreases exponentially with frequency, and the available thermal-noise power from a resistor *not limited* in bandwidth by any circuit element therefore is bounded, since this quantity is

$$N_t = kT \int_0^\infty p(f)\,df \tag{21}$$

In practice, circuit elements generally limit N_t before quantum effects, but it is comforting to know that a resistor cannot generate infinite noise power, as would be the case if $p(f)$ were absent from (21). Nature provides her own cutoff phenomena, a situation which limits the noise power traveling down any transmission line to a finite value.

In evaluating the thermal-noise power in a communication system, we need only include $p(f)$ if hf/kT is of the order of unity or greater. At a standard temperature of 290°K, $hf/kT = 1$ at 6,000 Gc, or a wavelength of 0.05 mm. Therefore, at standard temperatures we need consider $p(f)$ different from unity only when dealing with masers or lasers at submillimeter wavelengths. However, in a cryogenic closed system, the input-noise temperature to a maser amplifier might be a few degrees Kelvin. At 5°K, for example, the quantity hf/kT becomes equal to unity at about 100 Gc, a frequency presently used for communication systems. For the next generation of communication systems which will be operating from submillimeter wavelengths down to optical wavelengths, the complete expression for thermal-noise power must be used. Even in present-day systems, it should sometimes be considered at the higher microwave frequencies.

In Sec. 7-2, the mean-square noise current was defined in terms of the mean available power per unit bandwidth as

$$\overline{di^2} = 4G(f)w(f)\,df \tag{2}$$

where the circuit characterized by this current was assumed to be matched to a load. Equation (2) holds for the more general case, however, when the circuit is *not* matched to the load.† Therefore, the noise characterization of any passive circuit which can be represented by a conductance shunted by a susceptance is accomplished by adding a thermal-noise-

† For proof of this statement, see W. R. Bennett, "Electrical Noise," chap. 2, McGraw-Hill Book Company, New York, 1965.

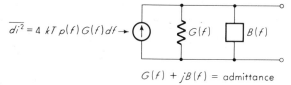

$G(f) + jB(f)$ = admittance

Fig. 7-7 Noise-current generator.

current generator of magnitude

$$\overline{di^2} = 4kTG(f)\,df \tag{22}$$

as shown in Fig. 7-7.

The same reasoning applies to the voltage-generator characterization, as shown in Fig. 7-8. If the example of Sec. 7-2 had been carried through with the conductance G in Fig. 7-4 as a function of frequency and not matched to the box, the mean-square output voltage obtained in (7) would be modified as follows:

$$\overline{e_o^2} = kT \int_0^\infty |Y(f)|^2 p(f) G(f)\,df \tag{7a}$$

EXERCISES

7-1a Verify Eq. (7a).

b Consider a series RLC circuit. Determine the energy stored in the capacitance from thermodynamic considerations and the average mean-square voltage across the capacitance.

c Calculate the mean-square output voltage for a parallel RLC circuit. *Hint:* To evaluate the integral, let $f/f_0 = e^x$, where f_0 is the resonant frequency.

7-2 Conductances g_1 and g_2 are in parallel and are both at temperature T. Determine the noise power in a bandwidth B delivered by these to a load conductance g_L by one of the following methods:

a Use one noise-current generator corresponding to the two conductances in parallel.

Fig. 7-8 Noise-voltage generator.

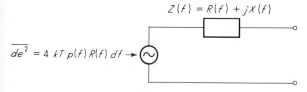

noise

b Use one noise-voltage generator corresponding to the two resistances in parallel.
c Use separate (incoherent) noise-current generators for g_1 and g_2.
d Use separate noise-voltage generators for the two.

7-3 A conductance g is in parallel with a capacitance C. What mean-square noise voltage in a bandwidth df at angular frequency would be observed by a high-impedance measuring device? Integrate to show that the total energy stored in the capacitance for all frequencies is $\frac{1}{2}kT$.

7-4 An antenna has induced in it a signal voltage e_0 (rms volts). It also receives noise which may be expressed in terms of an equivalent noise temperature T_a of its internal resistances R_r. The antenna is matched to the input of a tube of input resistance R_1 by an ideal transformer of turns ratio n. Shot noise in the tube may be expressed in terms of an equivalent noise resistance R_{eq} in its input circuit at room temperature T. Thermal noise from R_1 may also be computed at temperature T.
a Find the ratio of rms noise to rms signal voltage at the grid of the tube.
b For $T_a = T$ and $R_1/R_{eq} = 20$, plot this noise-to-signal ratio as a function of $n^2 R_r/R_1$.
c Why is it not a minimum at the turns ratio for impedance-matching R_r and R_1?

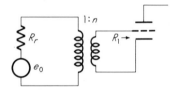

7-4. SHOT NOISE[2]

Electrical current is composed of a very large number of discrete charge carriers. Shot noise results when this number of carriers, present in some small volume or emitted from some surface, fluctuates randomly in time. The number of electrons leaving the cathode of a temperature-limited thermionic diode exhibits just such a random fluctuation, and as a result, the current flowing in a circuit containing the diode is correctly described as a time-invariant current (assuming the anode voltage and cathode temperature remain constant) plus a mean-square noise current, which is called the shot current. Because all electrons emitted by the cathode contribute to the current in a temperature-limited diode, the noise produced by this current is called *full shot noise*. All the electrons emitted by a thermionic cathode under space-charge–limited conditions do not reach the anode; if a larger-than-average number of electrons are emitted in a small interval of time dt, the potential minimum is depressed slightly and a slightly larger fraction of the emitted electrons are returned to the cathode. Similarly, if a smaller-than-average number of electrons are emitted during a small interval of time dt, the potential minimum is raised slightly and a larger-than-usual fraction of the emitted electrons proceed to the anode. The process described above effectively smooths the current in intervals of time long compared with the transit time between cathode and potential minimum, and thus reduces the current fluctuations at frequencies considerably less than the reciprocal of the cathode-to-potential-minimum transit time. Thus the noise current from a space-charge–limited cathode is substantially less than that from a temperature-limited cathode at low frequencies.

The noise current at higher frequencies, such as the microwave region, has received extensive study and is reasonably well understood in the region extending from slightly past the potential minimum to the anode. The region between the cathode and potential minimum is currently under active investigation. A detailed account of noise processes in microwave tubes is far beyond the scope of this work. An insight into the mathematical methods which prove useful in dealing with shot noise (and other types of noise as well) will instead be attempted. Since we are striving for physical understanding, not mathematical complexity, we shall not be rigorous here. The results we obtain, however, can be derived rigorously, and interested readers will be referred to the appropriate references.

First, we shall examine a "basic noise event." This event might be the current induced in an external circuit by the transit of an electron from the cathode to the anode of a thermionic diode, or it might be any other similar basic event. In the temperature-limited case, the current induced in the anode as a function of time might be given by the graph of Fig. 7-9. The frequency spectrum of this event is related to the time function by a pair of Fourier integral transforms as follows:

$$i(t) = \int_{-\infty}^{\infty} G(f)e^{j2\pi ft}\,df \tag{23}$$

$$G(f) = \int_{-\infty}^{\infty} i(t)e^{-j2\pi ft}\,dt \tag{24}$$

Parseval's theorem of Fourier analysis (see, for example, Goldman[3])

Fig. 7-9 Basic noise impulse.

states that

$$\int_{-\infty}^{\infty} i(t)i^*(t)\,dt = \int_{-\infty}^{\infty} G(f)G^*(f)\,df \qquad (25)$$

The left-hand side of the last equation may be interpreted physically as the energy which is dissipated in a 1-ohm resistor when a current pulse $i(t)$ passes through the resistor. If n of these pulses passed through the 1-ohm resistor each second and each pulse was completely independent of all the others (i.e., the pulses occur randomly spaced in time), the total energy dissipated per second is the sum of the energies dissipated per pulse, or

$$n\int_{-\infty}^{\infty} G(f)G^*(f)\,df = 2n\int_{0}^{\infty} G(f)G^*(f)\,df \qquad (26)$$

The power spectrum $w_n(f)$ is defined as

$$w_n(f) = 2nG(f)G^*(f) \qquad (27)$$

and, as indicated in the following paragraph, the mean-square value of the noise due to the n random pulses per second can be determined by integrating $w_n(f)$ over all positive frequencies. The power spectrum $w_n(f)$ is therefore interpreted as the mean-square noise contribution per unit bandwidth in the vicinity of a particular frequency f.

Rice[4] has used a more rigorous analysis to show the above. His formulation states that if a signal $i_n(t)$ is composed of like events $i(t)$ occurring randomly at an average rate of n per second,

$$\lim \frac{1}{T}\int_{0}^{T} i_n(t)i_n^*(t)\,dt = \int_{0}^{\infty} w_n(f)\,df \qquad (28)$$

where $w_n(f)$ is defined by (27). In the above equation, the left-hand side is interpreted as the time-average value of $|i_n(t)|^2$ or $\overline{i_n^2}$, taken over a very long time. The right-hand side is a different (and generally more convenient) method of calculating $\overline{i_n^2}$. It should be noted that the power spectrum $w_n(f)$ does not necessarily have the dimensions of power, and it should not be confused with the available mean power per unit bandwidth $w(f)$ introduced earlier, although, as we shall shortly see, it may be related to this quantity.

We have now a method of determining the mean-square value of a series of like, independent physical events occurring randomly in time. First, calculate the spectrum of one event. Second, multiply the spectrum by its complex conjugate and by the number of events occurring per unit time (which is assumed large). Next, integrate this quantity over the frequency band of interest, which is generally set by the measuring apparatus, and the resulting value is the value of the series of events averaged over many, many events, or in other words, over a very long time.

A special case of considerable importance in the study of full shot noise results when the transit time of the electron is very small compared with the reciprocal bandwidth of the instrument measuring the noise. In this case, the triangular pulse of Fig. 7-9 may be replaced by a properly weighted delta function whose properties are summarized as follows:

$$\delta(t - t_0) = 0 \quad \text{for } t \neq t_0 \tag{29}$$

$$\int_{-\infty}^{\infty} \delta(t - t_0)\,dt = 1$$

so that

$$\int_{-\infty}^{\infty} f(t)\delta(t - t_0)\,dt = f(t_0) \tag{30}$$

Since the area of the current pulse of Fig. 7-9 is the magnitude of the electronic charge e, we may write

$$i(t) = e\delta(t - t_0) \tag{31}$$

as a suitable approximation for the pulse when the transit time is short compared with any time of interest. (Alternatively, when the transit time τ is much shorter than $1/f_{\max}$, where $f_{\max}$ is the maximum frequency of interest, our approximation is valid.) Then

$$G(f) = e \exp(-j2\pi f t_0) \tag{32}$$

and

$$w_n(f) = 2ne^2 \tag{33}$$

But the number of electrons per second which pass from cathode to anode in any diode is $n = I_0/e$, so

$$w_n(f) = 2eI_0 \tag{34}$$

and

$$\overline{i_n^2} = 2eI_0 B \tag{35}$$

where B is determined in a practical case by the procedure of Sec. 7-2.

When the diode is space-charge–limited, at low frequencies the mean-square noise current can be written as

$$\overline{i_n^2} = \Gamma^2 2eI_0 B \tag{36}$$

where

$$\Gamma^2 = \frac{3(4-\pi)kTg}{2eI_0} \tag{37}$$

is the space-charge reduction factor and depends on the randomness ($\Gamma = 1$ for complete randomness) and where g is the ac conductance of the diode.

At high frequencies, the noise expressions are more complex.[5-7] In microwave tubes utilizing transit-time effects, such as klystrons and traveling-wave tubes, the noise is analyzed in terms of space-charge waves which propagate along the beam and are excited by both the shot current and the velocity variations of the electrons emitted from the cathode. Space does not permit a treatment of these relatively complicated phenomena.

EXERCISES

7-5 Show that the spectrum of $\delta(t - t_0)$ has a constant amplitude at all frequencies.

7-6 Show that $\delta(t - t_0)$ may be obtained by a limiting process from a rectangular pulse centered at $t = t_0$ of height h and width Δ. Let $h \to \infty$ and $\Delta \to 0$ such that $h\Delta = $ constant. Do the spectra of the δ function and the pulse (in the above limit) agree?

7-7 Determine the mean-square noise current from a temperature-limited diode measured by a suitable meter with an infinite bandwidth. If $\tau = 10^{-8}$ sec, estimate the mean-square noise current read by a meter which passes dc to 20 Mc, dc to 2 Gc, and dc to 20 Gc.

7-8 A resistor R at room temperature is connected in series with a temperature-limited diode so that the dc current I_0 passes through R. Under these conditions, the effective fluctuation current is twice that flowing through the diode alone. What is the direct voltage across R? Recall that the fluctuation current in a temperature-limited diode is given by

$$\overline{i_n^2} = 2eI_0B$$

and that

Room temperature = 300°K
$$k = 1.37 \times 10^{-23} \text{ joule/°K}$$
$$e = 1.6 \times 10^{-19} \text{ coul}$$

7-9 In a certain type of noise, individual bursts are triangular functions of time, as follows:

$$f(t) = Kt \quad 0 < t < \tau$$

If there are N of these bursts per second, occurring randomly, find the frequency spectrum of the noise power and sketch the results as functions of $\omega\tau$.

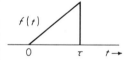

7-5. AMPLITUDE CHARACTERISTICS OF WHITE NOISE

In the discussions of noise so far we have talked mainly of the mean-square value of a particular fluctuation. It is of some interest, particularly in relation to many of the modulation schemes, to ask about the peak values of the noise wave (Fig. 7-10). Clearly this can be predicted only statistically, and Landon[8] has given a discussion which is useful at least for the general features of the phenomenon. He has

Fig. 7-10 Representation of noise wave.

derived the result

$$p = \frac{1}{2}\left[1 - \mathrm{erf}\left(\frac{V}{E2^{\frac{1}{2}}}\right)\right] \tag{38}$$

In the above, p is the probability that the noise voltage at any instant will exceed the value V and E is the rms value of the noise.

A curve representing (38) is plotted in Fig. 7-11. It should be noted from this that the probability of exceeding 0 amplitude is 0.5 (since the voltage will be negative half the time). The probability of exceeding the rms value in a positive sense is about 0.2, that of exceeding twice the rms value about 0.02, that of exceeding three times rms value about 0.001, etc. Although it is possible in principle to find peaks of arbitrarily high amplitude, it is clear that the probability becomes very small of finding peaks more than a few times the rms value. The mean value of the absolute voltage is 0.798 from this distribution.

Fig. 7-11 Probabilities p of amplitudes V where E = rms voltages.

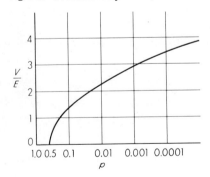

206
microwave communications

7-6. NOISE FIGURE AND RELATED TOPICS

When deciding which amplifier to choose in a given application where the noise performance is critical, it is useful to have a figure of merit which describes the noise performance of any amplifier on a common scale. Such a figure of merit is the noise figure, introduced by Friis[9] in 1944 and now widely used throughout the engineering profession. Consider the amplifier of Fig. 7-12. Let it be matched to the signal source at its input and to its output load. The available noise power per unit bandwidth at the amplifier input is kT, where T is the absolute temperature of Z_s. Generally, the amplifier itself generates noise, and this plus the amplified input-noise power is present at the output. If the output-noise power per unit bandwidth is $w_o(f)$ and the noise power per unit bandwidth generated by the amplifier referred to the input is $w_a(f)$, then

$$w_o(f) = A(f)kT + A(f)w_a(f) \tag{39}$$

The *spot noise figure* $F(f)$ of an amplifier is the available output-noise power per unit bandwidth divided by that portion of the output-noise power due to a matched input termination at the standard noise temperature $T_0 = 290°K$. Therefore

$$F(f) = 1 + \frac{w_a(f)}{kT_0} \tag{40}$$

If the output noise in a bandwidth B is N_o, then

$$N_o = \int_0^\infty w_o(f)\,df = A_o kTB + A_o N_a \tag{41}$$

where

$$A_o N_a = \int_0^\infty A(f)w_a(f)\,df \tag{42}$$

and is the output noise power due to the amplifier itself.

Fig. 7-12 Matched amplifier.

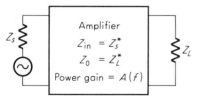

The average noise figure $\bar{F}$ of an amplifier over a bandwidth B is the total output-noise power divided by that portion of the output-noise power which originates in the matched input termination at the standard noise temperature of $T_0 = 290°K$. Thus

$$\bar{F} = 1 + \frac{N_a}{kT_0 B} = \frac{\int_0^\infty F(f) A(f)\, df}{\int_0^\infty A(f)\, df} \tag{43}$$

It is important to note that N_a, the noise power generated by the amplifier and referred to the input, could not be *measured* at the input; the measurable quantity is $A(f)w(f)\, df$ over some band of frequencies. In practice, the average noise figure is always measured; however, when the gain is different from zero over a sufficiently small bandwidth, the average and spot noise figures approach each other. A perfect amplifier would have a noise figure of unity, since $w_a(f)$ and N_a would vanish. Let us now calculate the output noise from an amplifier whose noise figure is given but whose input termination is not at temperature T_0, but at temperature T. From (43), we have

$$N_a = (\bar{F} - 1) k T_0 B \tag{43a}$$

and substituting this into Eq. (41), we see

$$N_o = A_o k B [T + (\bar{F} - 1) T_0] \tag{44}$$

If the input termination were cooled to 0°K, the entire noise output would be due to the amplifier. The magnitude of this output-noise power would be the same as if the input were matched to a source at temperature $(\bar{F} - 1)T_0$, the *excess noise figure* times the standard noise temperature. This temperature is called the average effective noise temperature T_e of the amplifier and is a very useful parameter in calculations dealing with low-noise amplifiers. Thus

$$T_e = T_0(\bar{F} - 1) \tag{45}$$

It is apparent that knowledge of T is equivalent to knowing the average noise figure, and if one is known, the other can be quickly calculated. Similarly, an effective spot noise temperature can be defined from (40).

The output-noise power of any amplifier is simply

$$N_o = A_o k (T_e + T) B \qquad (44a)$$

where T is the source temperature.

Often it is necessary to use several amplifiers in cascade, one following the other. If the amplifiers are matched and have spot noise figures F_i, a cascade of three amplifiers, as shown in Fig. 7-13 with an input termination at T_0, has an output-noise power per unit bandwidth of

$$w_o(f) = A_1 A_2 A_3 k T_0 + A_1 A_2 A_3 k T_{e1} + A_2 A_3 k T_{e2} + A_3 k T_{e3} \qquad (46)$$

and the portion of $w_o(f)$ due to the thermal-noise input is $A_1 A_2 A_3 k T_0$. The noise figure of the cascade may be written in terms of the individual noise figures and gains as

$$F = F_1 + \frac{F_2 - 1}{A_1} + \frac{F_3 - 1}{A_1 A_2} \qquad (47)$$

and the noise temperature of the cascade is

$$T = T_{e1} + \frac{T_{e2}}{A_1} + \frac{T_{e3}}{A_1 A_2} \qquad (48)$$

The analogous equations for the average figure and average effective noise temperature are more complicated unless the bandwidth of the last stage is much narrower than all preceding stages.

Consider now this design problem: Given two different low-noise amplifiers with noise figures F_1 and F_2 and gains A_1 and A_2, respectively, decide which should be used first in a cascade to give the lower overall noise figure. The noise figure with amplifier 1 used first is

$$F_{12} = F_1 + \frac{F_2 - 1}{A_1} \qquad (49a)$$

Fig. 7-13 Combined noise figure of cascade of three amplifiers.

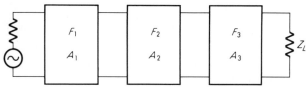

and the noise figure with the amplifiers reversed so amplifier 2 is first is

$$F_{21} = F_2 + \frac{F_1 - 1}{A_1} \tag{49b}$$

Let us determine the necessary condition that $F_{12} < F_{21}$. We find $(F_{12} - 1) < (F_{21} - 1)$, or

$$(F_1 - 1) + \frac{F_2 - 1}{A_1} < (F_2 - 1) + \frac{F_1 - 1}{A_2}$$

By algebraically manipulating the equation to get all the quantities which refer to amplifier 1 on the left side of the equation, we find

$$\frac{F_1 - 1}{1 - 1/A_1} < \frac{F_2 - 1}{1 - 1/A_2} \tag{50}$$

The quantity $(F - 1)/(1 - 1/A)$ has been called the *noise measure* by Haus and Adler.[10] For a cascade of two amplifiers to have the lower noise figure, the amplifier with the lower noise measure should come first. For a high-gain amplifier, the noise measure is merely the excess noise figure. For a low-gain amplifier, the noise measure may be considerably greater than the excess noise figure.

Loss often appears in a communication receiver, and in determining the noise figure, it must be included. We now calculate the noise figure of an attenuator at temperature $T_0 = 290°K$ with loss $L = P_{in}/P_{out} > 1$. The decibel loss is $10 \log L$, as usual. The input thermal-noise power per unit bandwidth from a matched source at $T_0 = 290°K$ is kT_0, and the noise-output power per unit bandwidth due to this input-noise power is kT_0/L. The total noise-output power per unit bandwidth must be kT_0, however, since an observer at the output looking toward the input sees a matched source at 290°K. Therefore

$$F = \frac{kT_0}{kT_0/L} = L \tag{51}$$

The noise figure of an attenuator at 290°K is also its loss.

EXERCISES

7-10 Derive the average noise figure for a cascade of two amplifiers if the bandwidths of both stages are comparable. The answer is

$$\bar{F} = \bar{F}_1 + \frac{\int_0^\infty A_2(F_2 - 1)\, df}{\int_0^\infty A_1 A_2\, df} \tag{52}$$

Show that this reduces to (47) when the bandwidth of amplifier 2 is much less than that of amplifier 1, but does not reduce to (47) if the reverse is true.

7-11 Show that if the attenuator is at an arbitrary temperature T, the noise power per unit bandwidth delivered to an external load is $kT[(L-1)/L]$.

7-12 A video amplifier has a bandwidth of 10 Mc and an input resistance (at room temperature) of 100 ohms.

a Assuming that the amplifier is noise-free, what is the available power gain required for an output-noise signal of 100 μv developed across a 300-ohm load?

b If the output observed in practice is 300 μv rather than 100 μv, what is the noise figure of the amplifier?

7-7. ANTENNA NOISE

One of the most important parts of a communication system is the antenna. Chapter 5 deals with the reception and transmission properties of the antenna when considering useful signals. It is convenient now to introduce the antenna as a receptor of noise.

Referring to (20), the thermal-noise power per unit bandwidth generated by a matched load at $T°K$ is given by

$$w(f) = kTp(f) \approx kT \tag{53}$$

For a perfectly matched antenna with no dissipative losses, the thermal-noise power per unit bandwidth delivered to the load would be given by

$$w(f) = kT_a \tag{54}$$

where T_a is the effective antenna-noise temperature.[11]

Assuming that the environment at which the antenna looks behaves according to the Rayleigh-Jeans radiation law, the blackbody brightness is given by

$$B(f) = \frac{2kf^2T}{c^2} \quad \text{watts}/(\text{m}^2)(\text{sr})(\text{bandwidth}) \tag{55}$$

where c is the velocity of light. The brightness relates the blackbody radiation per bandwidth at a temperature T and is randomly polarized. Since an antenna receives only one of any two polarization modes, then

$$B(f) = \frac{kT}{\lambda^2} \tag{56}$$

Consider the antenna having an effective receiving area A_e and being immersed in the region having the brightness (56). Then the total power per unit bandwidth received by the antenna is an integration over all space. Thus

$$w(f) = \int_\Omega \frac{kT}{\lambda^2} A_e \, d\Omega \quad \text{watts/bandwidth} \tag{57}$$

Here the noise arrives from the entire surrounding region represented by the total solid angle Ω and reaches the load through the effective receiving area A_e of the antenna. We saw that the effective area for an antenna is related to the gain g by

$$A_e(\theta,\phi) = \frac{\lambda^2}{4\pi} g(\theta,\phi) \tag{58}$$

and therefore (57) becomes

$$w(f) = \frac{k}{4\pi} \int_\Omega Tg \, d\Omega \tag{59}$$

Comparing with (54), we therefore define the effective noise temperature as

$$T_a = \frac{1}{4\pi} \int_{4\pi} T(\Omega)g(\Omega) \, d\Omega \tag{60}$$

Suppose that an antenna is completely omnidirectional [$g(\Omega) = 1$] and that the temperature surrounding it is $T(\Omega) = T_0$; then the antenna temperature will be

$$T_a = \frac{1}{4\pi} \int_{4\pi} T_0 \, d\Omega = T_0 \tag{61}$$

Now, on the other hand, suppose that the antenna sees a source of extent Ω_s with a temperature T_s, and suppose that the antenna pattern is contained in $\Omega_a > \Omega_s$, so $g \approx 4\pi/\Omega_a$. Then the effective antenna temperature will be

$$T_a = \frac{1}{4\pi} \int_{\Omega_s} T_s \frac{4\pi}{\Omega_a} \, d\Omega \approx T_s \frac{\Omega_s}{\Omega_a} \approx T_s \frac{g}{4\pi} \Omega_s \tag{62}$$

7-8. COMMUNICATION SYSTEM NOISE MODEL

The concepts of the previous sections can be summarized by considering the model[12] of a communication system of Fig. 7-14. The model consists of lossy passive networks at appropriate temperatures and an active network which is the receiver. The side-lobe noise power accounts for the terrestrial temperature effect received through the side lobes of the antenna. If the main beam of the antenna is pointing into a cold region and if the side lobes receive energy of about 300°K, the effective antenna temperature may be greatly affected.

The overall effective noise temperature referred to the input to the receiver is

$$(T_{e1})_{\text{eff}} = (T_{e1})_R + \frac{T_g}{L_1 L_2 L_3} + \frac{T_1(L_1 - 1)}{L_1 L_2 L_3} + \frac{T_2(L_2 - 1)}{L_2 L_3} + \frac{T_3(L_3 - 1)}{L_3} + \frac{T_{SL}}{L_3} \tag{63}$$

Fig. 7-14 Communication system model.

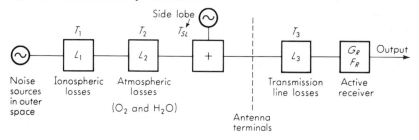

EXERCISES

7-13 Derive the expression in Eq. (63).

7-14 Given that the overall noise figure for two systems in cascade is

$$F = F_1 + \frac{F_2 - 1}{G_1}$$

where F_1 and F_2 are the noise figures of the two systems and G_1 is the power gain of the first, find the overall effective noise temperature of a transmission line of loss L at an actual temperature T_0, matched to a receiver of effective noise temperature T_R. Briefly explain the phenomenon of atmospheric absorption noise, and say why it is preferable to make radar astronomy observations vertically upward when operating in the 1-cm-wavelength range.

REFERENCES

1 IRE Standards on Circuits, *Proc. IRE,* vol. 48, pp. 1611–1612, September, 1960.

2 SCHOTTKY, W.: Über spontane Stromschwankungen in verschiedenen Elektrizitätsleitern, *Ann. Physik,* vol. 57, pp. 541–567, 1918.

3 GOLDMAN, S.: "Frequency Analysis, Modulation and Noise," chaps. 6–9, McGraw-Hill Book Company, New York, 1948.

4 RICE, S. O.: A Mathematical Theory of Random Noise, *Bell System Tech. J.,* vol. 23, pp. 282–332, July, 1944; vol. 24, pp. 46–156, January, 1945.

5 PIERCE, J. R.: The Physical Sources of Noise, *Proc. IRE,* vol. 44, pp. 601–608, May, 1956.

6 SCHWARTZ, M.: "Information Transmission, Modulation, and Noise," McGraw-Hill Book Company, New York, 1959.

7 BENNETT, W. R.: "Electrical Noise," McGraw-Hill Book Company, New York, 1960.

8 LANDON, V.: Distribution of Amplitude with Time in Fluctuation Noise, *Proc. IRE,* vol. 29, p. 50, February, 1941. (Discussions of this paper appeared in *Proc. IRE,* September, 1942, and November, 1942.)

9 FRIIS, H. T.: Noise Figures of Radio Receivers, *Proc. IRE,* vol. 32, pp. 419–422, July, 1944.

10 HAUS, H. A., and R. B. ADLER: "Circuit Theory of Linear Noisy Networks," The Technology Press of the Massachusetts Institute of Technology and John Wiley & Sons, Inc., New York, 1959.

11 BALAKRISHNAN, A. V. (ed.): "Space Communications," p. 145, McGraw-Hill Book Company, New York, 1963.

12 DEROSA, L. A., and E. W. KELLER: Potential Application of Recent Advances in Communication Technology, *Elect. Commun.,* vol. 37, no. 3, 1962.

8

specific microwave communication systems

8-1. INTRODUCTION

As was stated in Chap. 1, the main problem of a new system is to properly design and assemble it from its independent elements for the purpose of obtaining a common objective. How this is done cannot be easily outlined. There needs to be a combination of experience and knowledge of new techniques, devices, and components as well as satisfaction of the requirements imposed by the constraints of the problem.

A major problem arising in the final design of a system is that of ultimately selecting the optimum system. This choice may depend on physical limitations (size, weight, etc.), economic considerations, or technical circumstances set by either nature or the customer. There exist many references which deal with the details of system designing. A few of these are listed at the end of the chapter. It is beyond the scope of this text to present anything more than a simple illustration of the problem encountered in a design.

8-2. A RADAR SYSTEM

In order to illustrate how the concepts and devices discussed in the previous chapters of this text are interrelated, a known system, namely, a radar, is now considered. The word *radar* is a contraction of *r*adio *d*etection *a*nd *r*anging. The elementary form of a radar consists primarily of a transmitter and a receiver. The transmitter emits electromagnetic radiation generated by an oscillator. The electromagnetic energy is "scattered" in the back direction and is an important element in radar. The receiver detects and processes the scattered energy, thus determining the presence, location, and velocity of the target. This, in essence, is a communication system (see Chap. 1). The information on distance is obtained by a measure of time taken for the radar wave to reach the target and to return. Directional information is obtained by means of the narrow antenna beam and from the arrival characteristics of the reflected wave. The radial velocity of the target is determined through a doppler shift in the carrier frequency. Another measure of velocity is obtained by continuous monitoring of the change of the observed position of the target.

A block diagram of a simplified radar system is shown in Fig. 8-1. The timing circuit supplies a low-frequency pulse train. Typically, the pulse width τ ranges from 0.1 to 10 μsec, and the pulse repetition rate f_r ranges from 250 to 500 pps. Duplexers consist of special transmit-receive switches (TR and ATR devices) which, among other things,

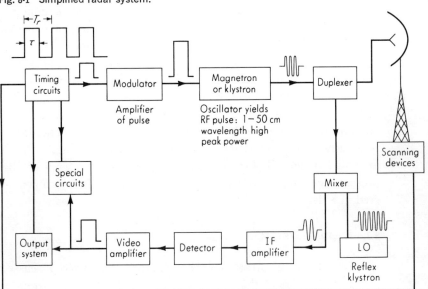

Fig. 8-1 Simplified radar system.

allow for receiver protection during the time of transmission. When it is active, the ATR permits the generated pulse to pass through to the antenna; when it is quiescent, it blocks the path of the received pulse toward the signal generator in order to avoid reflection problems. When struck, the TR shorts the transmitted pulse from the mixer and detector; when it is quiescent, it allows the weak pulse to pass to the receiver. Many special circuits may appear even in a simplified version of a radar. These circuits may consist of gating circuits which time-separate the received pulse from the transmitted pulse, duplexers and ferrite devices, etc.

Choice of Frequency

A possible starting point for the design of the radar system may be the choice of frequency of operation. This is a reasonable starting quantity since the wavelength is involved implicitly in almost every radar parameter. The frequency affects the character of the propagation path and radar environment. Some radar parameters are best suited for lower frequency, e.g., weather effects, atmospheric attenuation, and ease of generating large power. Other parameters are more readily obtained at the higher frequencies, e.g., narrow antenna beamwidths and freedom from cosmic noise. On the other hand, some parameters are relatively independent of frequency, e.g., receiver bandwidth, pulse repetition frequencies, and signal-to-noise ratio. As is seen in Fig. 6-11, the *atmospheric losses* vary considerably with frequency. The frequency range of 18 to 26.5 Gc (K band) centers almost directly on the H_2O absorption line at 22.3 Gc. It is seen that neither the 3- nor the 10-Gc-frequency regions experience formidable atmospheric losses. In fog or rain the losses for visibility range of 400 ft are shown in the table below.

LOSS, db/km	WAVELENGTH, cm
0.035	> 6
0.35	≈ 3
1–3	≈ 1

In clear weather there is little absorption for wavelengths as small as 2 cm, after which the attenuation is as shown in the table below.

LOSS, db/km	WAVELENGTH
0.25	≈ 1.4 cm
15	≈ 5 mm, 60 Gc

Another parameter which can influence the choice of frequency is the noise arising from *cosmic* sources. Below is a summary of the relationship between the source of noise and its dependence on wavelength.

SOURCE	FLUX DENSITY	ANTENNA TEMPERATURE	REMARKS
Moon or earth looking down	$\propto \lambda^{-2}$	$N = \int_\Omega \dfrac{kT}{\lambda^2} \dfrac{\lambda^2}{4\pi} g\, d\Omega \propto \lambda^0$	Typical blackbody
Sun	$\propto \lambda^{-0.75}$	$N = \int_\Omega A\lambda^{-0.75} \dfrac{\lambda^2}{4\pi} g\, d\Omega \propto \lambda^{1.25}$	Under normal conditions, no excess sun spots
Radio stars	$\propto \lambda^{0.5}$	$N = \int_\Omega A\lambda^{0.5} \underbrace{\dfrac{\lambda^2}{4\pi}}_{A_e} g\, d\Omega \propto \lambda^{2.5}$	Cygnus A, Taurus A

Range of Target

The range of a target is determined by the measured total time to and from the target. Thus

$$R = \frac{c\,\Delta t}{2}$$

where c is the velocity of the electromagnetic wave. If Δt equals 1 μsec, the range will be 150 m.

The Maximum Unambiguous Range

For the simple radar system under consideration, the maximum range is limited by the overlap of the received pulse by the transmitted pulse train of repetition frequency f_r. Hence

$$R_{\text{unambiguous}} = \frac{c}{2f_r} = \frac{cT_r}{2}$$

Radial Velocity of Target

A moving-target–indicating (MTI) radar determines the radial velocity by noting the apparent change in frequency of the returning pulse of oscillation. In the two-way path to the target and back to the radar the total number of wavelengths contained is $2R/\lambda$. Since a single wavelength corresponds to an angular excursion Φ of 2π rad, the total angle of the two-way path is

$$\Phi = 2\pi \frac{2R}{\lambda} = \frac{4\pi R}{\lambda}$$

When R is changing, so does Φ and the instantaneous angular frequency is

$$\omega_{\text{inst}} = \frac{d\Phi}{dt} = \frac{4\pi}{\lambda} \frac{dR}{dt} = \frac{4\pi v_r}{\lambda}$$

where v_r is the radial component of the relative velocity of the target. Hence the doppler frequency is

$$f_{\text{dop}} = \frac{2v_r}{\lambda} = \frac{2v_r f}{c}$$

and with

$$v_r = v \cos \theta$$

the doppler frequency is then

$$f_{\text{dop}} = \frac{2v(\cos \theta)f}{c}$$

Radar Equation

An important equation relating the power requirements and the characteristics of the antenna and target is called the radar equation.

specific microwave communication systems

Let W_t be the power of the radar transmitter. Then $W_t g_t/4\pi R^2$ represents the power density at a distance R from a transmitting antenna of gain g_t. Signify the power reradiated by the target toward the transmitter by $(W_t g_t/4\pi R^2)\sigma$, where σ is the radar cross section of the target and has the dimensions of area. It will be discussed in a later section. The factor $[W_t g_t/(4\pi R^2)^2]\sigma$ gives the power density of the echo signal at the radar. The radar receiving antenna of effective cross section A_e captures a portion of the echo power which is

$$W_r = \frac{W_t g_t \sigma}{(4\pi R^2)^2} A_e \tag{1}$$

This is the fundamental form of the radar equation. From Chap. 5 we have seen that the antenna parameters may be related by

$$g_t = \frac{4\pi A_t}{\lambda^2} \qquad g_r = \frac{4\pi A_r}{\lambda^2}$$

If the transmitting and receiving antennas are one and the same, the received power may be expressed as

$$W_r = \frac{W_t A_e \sigma}{4\pi \lambda^2 R^4}$$
$$W_r = \frac{W g^2 \lambda^2 \sigma}{(4\pi)^3 R^4} \tag{2}$$

The maximum radar range $R_{\max}$ is the distance beyond which the target can no longer be detected. It occurs when the received echo signal W_r just equals a minimum detectable signal $S_{\min}$, and therefore

$$R_{\max} = \left(\frac{W_t A_e^2 \sigma}{4\pi \lambda^2 S_{\min}}\right)^{\frac{1}{4}} = \left[\frac{W_t g^2 \lambda^2 \sigma}{(4\pi)^3 S_{\min}}\right]^{\frac{1}{4}} = \left[\frac{W_t g A_e \sigma}{(4\pi)^2 S_{\min}}\right]^{\frac{1}{4}} \tag{3}$$

Care must be taken in interpreting Eqs. (1) to (3) as to their dependence on wavelength since effective areas, radar cross section, and gain are themselves functions of the wavelength.

Radar Cross Section

The radar cross section σ is defined as the area intercepting the amount of power which when scattered isotropically produces an echo at the source of radiation equal to that observed from the target. In terms of E_r, the reflected field strength at the radar, and E_i, the strength of the incident field at the scatterer, which is a distance R from the radar, the cross section is

$$\sigma = 4\pi R^2 \frac{|E_r|^2}{|E_i|^2} \tag{4}$$

As an application of (4) consider any matched antenna. The power received in the matched load is

$$W_{\text{rec}} = \frac{E_i^2}{\eta} A_e \tag{5}$$

but since the antenna is matched, an equal power is lost in the radiation resistance. This is the reradiated, or scattered, power from which is obtained

$$E_r = \frac{\sqrt{30 W_{\text{rec}} g_t}}{R} \tag{6}$$

where g_t is the gain of the scattering antenna. The backscatter cross section of the matched antenna is, therefore,

$$\sigma = 4\pi R^2 \frac{30 g_t W_{\text{rec}}}{R^2 E_i^2} = 4\pi \frac{R^2 30 g_t}{R^2 E_i^2} \frac{E_i^2 A_e}{\eta} = g_t A_e = g_t^2 \frac{\lambda^2}{4\pi}$$

since $A_e = \lambda^2 g_t / 4\pi$. For a matched half-wave dipole

$$\sigma = (1.64)^2 \frac{\lambda^2}{4\pi} = 0.22 \lambda^2 \tag{7}$$

For a short-circuited half-wave antenna, that is, with the load replaced by the short circuit, the current induced would be twice that of a matched dipole and the rescattered field would be doubled so that the cross section would be four times that of a half-wave matched dipole.

The radar backscatter cross section of other scatterers whose dimensions are large compared with wavelength can be shown to be as in the table below.

specific microwave communication systems

SCATTERER	RADAR BACKSCATTER CROSS SECTION σ
Corner reflector with a diagonal of a	$\dfrac{4\pi a^4}{3\lambda^2}$
Flat plate area A	$\dfrac{4\pi A^2}{\lambda^2}$
Cylinder with radius a, length L	$\dfrac{2\pi L^2 a}{\lambda}$
Small plane	$20\ m^2$ (approx)

Methods of Calculation of Radar Cross Section

In order to determine the radar cross section exactly, we need to solve Maxwell's equations, subject to the boundary conditions on the

Fig. 8-2 Scattering from a sphere.

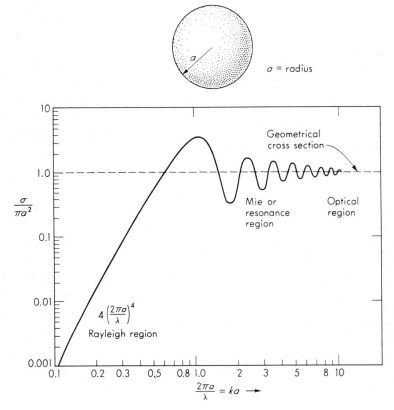

obstacle, matching the incident field with the scattered fields. Except for very simple shapes, e.g., a body having a contour surface of a coordinate system for which the scalar wave equation is separable, this is a difficult procedure. The sphere is one shape for which the analysis involves a reasonable amount of calculation. Figure 8-2 shows the variation of the backscattering by a sphere as a function of the ratio of the radius to the wavelength. Figure 8-3 indicates the complexity of the scattering cross section of an aircraft and that of a relatively simple object such as a truncated cone. The backscattering cross-section variation is represented on a polar diagram by the scattered electric field intensity. Obviously, approximate methods must be used to obtain even numerical answers as to the magnitude of the scattering cross section. For the *physical optics* case the wavelength of the incoming radiation must be small compared with the dimensions of the scattering body. The *geometrical optics* case is the limit case of physical optics having vanishingly small wavelengths. The *Rayleigh law* of scattering applies when the wavelength is much greater than the dimensions of the scattering body. The direct solution may be necessary when the wavelength of the radiation is of the same order as the dimensions of the scatterer.

To illustrate the method and approximations used in determining the Rayleigh scattering for objects small compared with wavelength, consider the following case. The local fields for a spherical conductor

Fig. 8-3 Backscattering. (a) Complex scatterer; (b) simple scatterer (truncated cone).

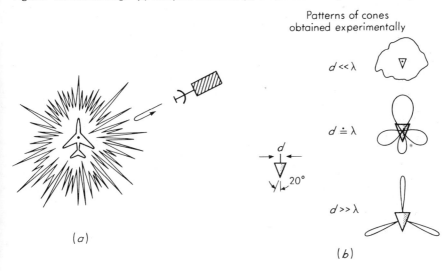

specific microwave communication systems

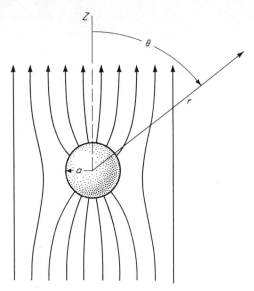

Fig. 8-4 Spherical conductor in a uniform field.

in a uniform field (Fig. 8-4) for an electrostatic case are obtained from the electrostatic potential

$$\Phi = \sum_{n=0}^{n=\infty} \left(A_n r^n + \frac{B_n}{r^{n+1}} \right) P_n \qquad (8)$$

where r is measured from the center of the sphere and where P_n represents the Legendre polynomials as follows:

$P_0 = 1$
$P_1 = \cos \theta$
$P_2 = \tfrac{1}{2}(\cos^2 \theta - 1)$
. . . .

Before the sphere is introduced, the potential of the uniform field may be taken to be

$$\Phi = -E_0 r \cos \theta$$

On the sphere, set $\Phi = 0$ at $r = a$, where a is the radius, and off the sphere for $r \gg a$, set $\Phi = -E_0 r \cos \theta$. Examination of the Legendre polynomials shows that only solutions having $n = 0$ and $n = 1$ can be expected to satisfy these conditions. Therefore, (8) reduces to

$$\Phi = A_0 + \frac{B_0}{r} + A_1 r \cos\theta + \frac{B_1}{r^2} \cos\theta \tag{9}$$

Applying the boundary conditions for $r \to \infty$ and $r = a$, we simplify (9) to

$$\Phi = -E_0 r \cos\theta + E_0 \frac{a^3}{r^2} \cos\theta \tag{10}$$

The electrostatic electric field is then

$$\mathbf{E} = -\nabla\phi = \left(E_0 \cos\theta + \frac{2E_0 a^3}{r^3} \cos\theta\right)\mathbf{a}_r + \left(-E_0 \sin\theta + E_0 \frac{a^3}{r^3} \sin\theta\right)\mathbf{a}_\theta$$

and further

$$\mathbf{E} = E_0 \mathbf{a}_z + E_0 \frac{a^3}{r^3}(\mathbf{a}_r\, 2\cos\theta + \mathbf{a}_\theta \sin\theta) \tag{11}$$

We now compare (11) with the field that arises from two charges of opposite sign, i.e., from an electrical dipole. From this comparison it will be possible to deduce the time-varying far-zone fields. Referring to Sec. 5-3, we see that the electric field of a dipole is

$$\mathbf{E} = \frac{p}{4\pi\epsilon_0 r^3}(\mathbf{a}_r\, 2\cos\theta + \mathbf{a}_\theta \sin\theta) \tag{12}$$

where $p = ql$ for the two point charges of magnitude $\pm q$ at a distance l apart. Summarizing, we see that the electrostatic electric fields are

1. For a sphere

$$\mathbf{E} = E_0 \frac{a^3}{r^3}(\mathbf{a}_r\, 2\cos\theta + \mathbf{a}_\theta \sin\theta)$$

2. For a dipole

$$\mathbf{E} = \frac{p}{4\pi\epsilon_0 r^3}(\mathbf{a}_r\, 2\cos\theta + \mathbf{a}_\theta \sin\theta)$$

$$\tag{13}$$

The electrostatic dipole moment for a sphere is deduced to be

$$p = 4\pi\epsilon_0 a^3 E_0 \tag{14}$$

Again from Sec. 5-3 for a *radiating dipole*, the electric field is

$$\mathbf{E} = \frac{p e^{j\omega(t-r/v)}}{4\pi\epsilon_0} \left[\frac{-\omega^2 \sin\theta}{v^2 r} \mathbf{a}_\theta + \left(j\frac{\omega}{vr^2} + \frac{1}{r^3} \right) (\mathbf{a}_r\, 2\cos\theta + \mathbf{a}_\theta \sin\theta) \right] \quad (15)$$

Note that for $r \ll \lambda$, $\mathbf{E}$ reduces to that for the electrostatic case and is also identical in form to that arising from the dipole property of the sphere. It is concluded that for large r the "reflected field" is

$$E_r \mathbf{a}_\theta = \frac{p}{4\pi\epsilon_0} \left(-\frac{\omega^2 \sin\theta}{v^2 r} \right) \mathbf{a}_\theta = \frac{4\pi\epsilon_0 a^3}{4\pi\epsilon_0} E_0 \left(-\frac{\omega^2 \sin\theta}{v^2 r} \right) \quad (16)$$

For backscattered fields, $\theta = 90°$ and $E_0 = E_i$ and therefore the reflected electric field intensity is

$$E_r = \frac{a^3 E_i}{r} \left(\frac{2\pi}{\lambda} \right)^2 \quad (17)$$

If we replace r by R in (17) and use (4), the scattering cross section, due to the electric dipole effect, becomes

$$\sigma_E = 4\pi R^2 \left| \frac{a^6 E_i^2}{R^2 E_i^2} \left(\frac{2\pi}{\lambda} \right)^4 \right| = 4\pi a^2 \left(\frac{2\pi a}{\lambda} \right)^4 \quad \text{meter}^2 \quad (18)$$

In this development, the magnetic dipole effect has been ignored. In fact, for small ratios of a/λ, the magnetic dipole coefficient becomes half as large as that for the electric dipole and must be included. When both electric and magnetic dipoles are considered, the backscattering cross section σ is $(\frac{3}{2})^2$ that of (18). The importance of wavelength is indicated by its inverse fourth-power dependence.

The basis for the physical and geometrical optics calculation is Kirchhoff-Huygens' principle (Sec. 5-7). Briefly stated, the principle is as follows: If the value of a field quantity is known at every point on a closed surface surrounding a source-free region, each elementary unit of surface can be considered as a radiating source.

The total field at any interior point is given by integrating the contributions of all elements over the surface. For the geometrical optics case the wavelength is vanishingly small. As an example of this method of calculation of the scattering cross section consider the physical optics case. The following conditions and restrictions are imposed:

a. Only backscattering is to be calculated.
b. The perfect conductor scatterer is considered.
c. The dimensions are much greater than the wavelength.
d. No sharp corners exist.
e. The distance of the scattered field from the scatterer is much greater than the dimensions of the scatterer.

Let the incident magnetic field intensity be

$$\mathbf{H}_i = H_0 e^{-jky} \mathbf{a}$$

Then the total magnetic field on a conducting surface is

$$\mathbf{H}_t = \mathbf{i}_t 2 H_0 e^{-jky}$$

where $\mathbf{i}_t$, the unit vector tangent to the surface, is related to the normal to the surface and field orientation by

$$\mathbf{i}_t = \mathbf{a} - (\mathbf{a} \cdot \mathbf{n})\mathbf{n}$$

Analogous to the expressions of Eqs. (59), Chap. 5, the scattered magnetic field intensity is

$$\mathbf{H}^s = -\frac{1}{4\pi} \int_S \left[(\mathbf{n} \times \mathbf{H}_t) \times \nabla \frac{e^{-jkr}}{r} \right] dS \tag{19}$$

with the approximation that

$$\nabla \left(\frac{e^{-jkr}}{r} \right) \approx -\mathbf{n}_0 \left(jk \frac{e^{-jkr}}{r} \right) \tag{20}$$

using the restrictions listed in (a), (c), and (e) above and with

$$\mathbf{n}_0 = -\mathbf{n}_i \tag{21}$$

where $\mathbf{n}_i$ is the direction of the incident electromagnetic wave. As is done in the usual physical optics approximation, r is removed from under the integral sign where it affects only the amplitude. Noting that $\mathbf{n}_0 \times (\mathbf{n} \times \mathbf{H}_t) = -2\mathbf{a}H_0 e^{-jky}(\mathbf{n}_0 \cdot \mathbf{n})$, we see that the scattered magnetic field becomes

$$\mathbf{H}^s = \frac{-jk}{4\pi r} \int_S \mathbf{n}_0 \times (\mathbf{n} \times \mathbf{H}_t) e^{-jkr}\, dS \tag{22}$$

where dS is the element of area on the scattering surface. Then dA may be considered as the projection of dS on the plane normal to the line of sight so that

$$dA = -(\mathbf{n}_0 \cdot \mathbf{n})\, dS \tag{23}$$

For backscattering

$$\mathbf{H}^s = \frac{-jkH_0\mathbf{a}}{2\pi r} \int_A e^{-2jky}\, dA \tag{24}$$

where y is the coordinate along the direction of propagation of the incident field. The scattering cross section in terms of the scattered power density P_s and incident power density P_i is

$$\sigma = 4\pi r^2 \frac{|P_s|}{|P_i|} = 4\pi r^2 \frac{|H^s|^2}{|H_0|^2} \tag{25}$$

and upon substitution of (24) in (25), we obtain

$$\sigma = \frac{4\pi}{\lambda^2} \left| \int_A e^{-2jky}\, dA \right|^2 = \frac{4\pi}{\lambda^2} |\gamma|^2 \tag{26}$$

where

$$\gamma = \int_A e^{-2jky}\, dA \tag{27}$$

Examples of the use of (26) are simple when the geometry is a simple one. For a flat plate it is easily seen that $\sigma = 4\pi A^2/\lambda^2$ since

$$|\gamma|^2 = A^2$$

where A is the area of the plate. The backscattering from a sphere in the geometrical optics case may also be easily obtained. Referring to Fig. 8-5, we see that $y = a - a \cos \psi$. Now (27) becomes for the sphere

$$\gamma = \int_0^{\pi/2} e^{-2jk(a - a\cos\psi)} \underbrace{a\, d\psi\, 2\pi a \sin \psi}_{dS} \cos \psi \tag{28}$$

Let $\zeta = \cos \psi$, so that $d\zeta = -\sin \psi\, d\psi$. Then (28) becomes

$$\gamma = e^{-2jka} 2\pi a^2 \int_0^1 e^{2jka\zeta} \zeta\, d\zeta \tag{29}$$

Upon integration and use of assumption (c) above, we obtain

$$|\gamma|^2 = \frac{\lambda^2 a^2}{4} \tag{30}$$

The sphere scattering cross section for $2\pi a/\lambda \gg 1$ becomes

$$\sigma = \frac{4\pi}{\lambda^2} |\gamma|^2 = \frac{4\pi}{\lambda^2} \frac{\lambda^2 a^2}{4} = \pi a^2 \tag{31}$$

Maximum Radar Range

Referring back to the maximum range (3), we see that

$$R_{\max} = \left[\frac{W_t g^2 \lambda^2 \sigma}{(4\pi)^3 S_{\min}} \right]^{\frac{1}{4}}$$

It is seen that the range is dependent on the choice and nature of several

Fig. 8-5 Geometry of spherical scatterer.

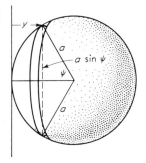

parameters: the minimum detectable signal S_{min} depends on (1) signal-to-noise ratio and (2) the false alarm ratio, i.e., the number of pulses of noise per second having the magnitude of order of that of the signal pulse or the number of intercepted pulses per target.

The signal-to-noise ratio may be written as (see Chap. 7)

$$\frac{W_s}{N} = \frac{W_s}{kT_nB} \qquad (32)$$

where T_n is an effective noise temperature. Assume that detection will be possible if the ratio

$$\frac{W_s}{kT_nB} \geq \chi \qquad (33)$$

which depends on assumptions (a), (b), and (c) above. Substituting

$$S_{min} = \chi kB(T_a + T_r) \qquad (34)$$

in Eq. (3) yields

$$R_{max} = \left[\frac{W_t q_a{}^2 \lambda^2 \sigma}{(4\pi)^3 kB(T_a + T_r)\chi} \right]^{\frac{1}{4}} \qquad (35)$$

where T_n has been replaced by the sum of the antenna and receiver noise temperatures. If, through great effort, the noise temperature of the receiver is reduced to zero, it is seen that the improvement is 16 percent over a hypothetical case of equal receiver and antenna temperatures.

For a one-way communication system, the maximum range may be easily shown to be

$$R_{max} = \left[\frac{W_t g_t A_R}{4\pi kB(T_a + T_r)\chi} \right]^{\frac{1}{2}} \qquad (36)$$

For this system a reduction of receiver temperature to zero yields an improvement to approximately 30 percent. In Eq. (35) the bandwidth B is an important parameter. If we consider the pulse train of Fig. 8-6 impressed on a filter of various bandwidths, the resulting pulse shape as a function of bandwidths[1] is shown by the solid lines of Fig. 8-7. Also

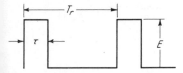

Fig. 8-6 Pulse train.

to be noted is that the noise power is proportional to the bandwidth. Figure 8-8 shows the variation of noise power, signal power, and signal-to-noise ratio as a function of the bandwidth. A maximum signal-to-noise ratio appears at about $B = 1.2/\tau$. Obviously, it is not advisable to continue to increase the bandwidth above this value unless use of leading edge of the pulses for position indication is required. The maximum range is related to the noise figure of the second detector and a visibility factor V as follows:

$$\chi = \tilde{F} V$$

where $\tilde{F}$, the noise figure before the second detector, is

$$\tilde{F} = L_{\text{RF}} \frac{1}{G_1} (t_r + F_{\text{IF}} - 1) \tag{37}$$

(see Chap. 7), t_r is the crystal excess noise, and L_{RF} is the loss at radio frequency. The visibility factor is a function of the probability of detection and depends on the pulse length, repetition rate, and other parameters. It is beyond the scope of this text to go into the details which determine its optimum value. It has been shown that the visibility factor is related to the pulse repetition frequency by the empirical expression

$$V = \sqrt{\frac{50}{f_r}} \tag{38}$$

Combining all the aforementioned parameters and substituting $B = 1.2/\tau$, we see that the maximum range becomes

$$R_{\max} = \left(\frac{W_t}{kT} \frac{A_e^2 \sigma}{1.2 \tilde{F} \sqrt{50/f_r} \, 4\pi\lambda^2} \right)^{\frac{1}{4}} \tag{39}$$

Note that the first fraction in the parentheses represents the ratio of

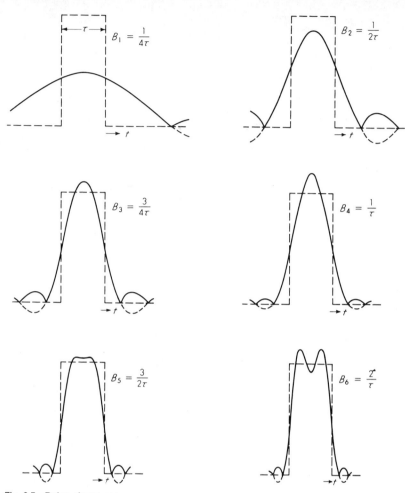

Fig. 8-7 Pulse shape response.

energies, i.e., the ratio of energy out of the transmitter to energy for one degree of freedom. If the peak pulse power is kept fixed and the pulse width is increased by a factor of 4, the maximum range will be increased by 41 percent. However, by so doing, we increase the energy in the pulse.

The Duty Cycle

The duty cycle of the train of pulses is $D = W_{av}/W_{peak} = \tau f_r = \tau/T$.

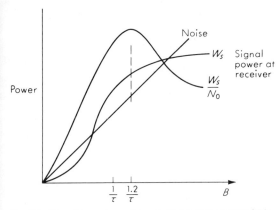

Fig. 8-8 Signal-to-noise ratio variation versus bandwidth.

The Pulse Width

Inspection of Eq. (39) suggests that τ should be chosen as large as possible within proper operation conditions of the signal generator. However, the choice of a large value of τ is limited by the following considerations.

The *range resolution* is better if short duration pulses are used. For example, if there are two targets at approximately the same azimuth and elevation, they cannot be distinguished if they are closer than the distance corresponding to approximately one-half the pulse width. In other words, for $d \leq c\tau/2$, where d is the radial distance between the pulses, the two pulses from the targets would blend into one. The coverage area is obscured by *ground clutter* which generally increases with increasing pulse width. If a *magnetron* is used as a source of pulsed radio frequency, its behavior may become critical as the pulse length increases.

The pulse duration has a significant effect on the intensity of *cloud return* relative to signal return from targets. Generally the undesired return varies directly with pulse duration as follows:

$$\frac{\text{Average rain echo intensity}}{\text{Target echo intensity}} = \left(\frac{R^2\lambda^2}{A}\frac{c\tau}{2}\right) N \frac{\sigma_0}{\sigma_t}$$

where R = range
λ = wavelength
c = velocity of light

τ = pulse duration
A = antenna aperture
N = number of drops per unit volume
σ_0 = radar cross section of an average drop = $4\pi(2\pi/x)^4 \bar{a}^6$
σ_t = radar cross section of target
$\bar{a}$ = average radius of drops

The scattering from rain drops may be reduced by making use of the symmetry of the water droplets. If a sphere is struck by a circularly polarized plane wave, the scattered wave observed at the source of the circularly polarized wave will be circularly polarized but in the opposite sense; and the receiving antenna may be designed to reject this backscattered wave. On the other hand, targets generally lack the symmetry of a water drop, so that the backscattered wave is elliptically polarized. Such a wave contains circular polarization of both types and can be easily detected.

Pulse Recurrence Frequency

The pulse recurrence frequency must be carefully chosen in order that the condition for the maximum unambiguous range mentioned earlier in this chapter be satisfied; namely, $1/f_r = T_r \geqq 2R/c$. If f_r is chosen to be 400 pps, the unambiguous range will be 375 km.

Scanning Rate

The scanning rate must be chosen after consideration of the fact that the scanning rate sets the lower limit of the pulse recurrence frequency. The limiting accuracy of the azimuthal angle determination is approximately half the angular separation between pulses. Let Θ be an average angle indication of where the target is. Then as the antenna beam scans, $\Theta = \omega_{sc}T_r$ and $\omega_{sc} = \Theta f_r$, where ω_{sc} is the scanning angular rate. The scanning rate is also related to the angular width of the radiation beam ($\theta°$ half-power width). Let N_{sc} be the number of pulses on a target during a scanning. Then $N_{sc} = (\theta°/\omega_{sc})/T_r$ = the ratio of the time the beam is on target to the time between pulses. Also

$$N_{sc} \geqq 1 \qquad \frac{f_r \theta}{\omega_{sc}} \geqq 1$$

If θ is taken to be 1° and f_r = 400 pps, then ω_{sc} = 1.11 rad/sec.

Summary

The above example is only an illustration of a simple microwave system. More in line with microwave communication systems are the examples contained in the following exercises. In addition, several of these exercises have been chosen to combine in various systems many of the concepts presented throughout the earlier chapters. Obviously, no unique system will result since so much depends on the choice of the parameters and components.

EXERCISES

8-1 Show the missing steps in the derivation for the backscattering cross section [Eq. (18)].

8-2 Derive the expression for the backscattering cross section for a sphere [Eqs. (29) to (31)].

8-3 A radar beacon utilizes a standard radar transmitter-receiver with output power W_1, noise figure F_1, and antenna gain G_1. The target here is a "transducer" which receives the radar pulse in a receiver of noise figure F_2 and, if the received signal is above a threshold minimum, reradiates from a second transmitter of power W_2 and antenna gain G_2 toward set 1.

In some detail, discuss the maximum range of the system.

The circulator (see figure) is a device which the energy circulates in a prescribed manner. Energy entering port 1 exits at port 2, energy entering port 2 exits through port 3, and energy entering port 3 exits at port 1. Generally, all the ports are matched to the external loads; i.e., there are no reflections between the circulator ports and the elements to which they are connected.

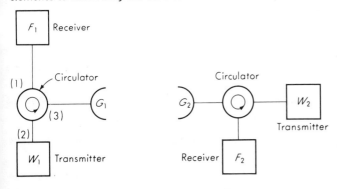

8-4 The following information is available regarding a microwave relay link used to communicate between the mainland and an island. It employs identical parabolic antennas at the transmitter (height h_T) and the receiver (height h_R).

In addition,

f = frequency

B = bandwidth

F = noise figure of receiver

$(S/N)_o$ = signal-to-noise power ratio needed at output of receiver

$A_e(\theta,\phi)$ = effective area of antennas

Develop a range equation for the microwave link, taking into account reflections.

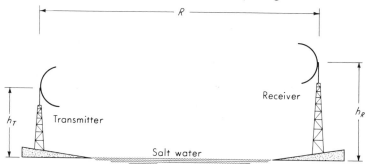

8-5 Assume you wish to use ham (amateur) radio to talk to Hawaii several evenings a week. Using references such as those listed below (and data from equipment catalogs) submit a design which has considered a suitable and allowable frequency to be used, transmission path and means, noise, transmitter power, polarization and antennas at both locations. Give sufficient detail to indicate reasons for your choice of design. The references are as follows:

"Reference Data Radio Engineers," 4th ed., chap. 24, International Telephone and Telegraph Corporation, New York, 1956.

F. E. Terman, "Electronic and Radio Engineering," 4th ed., chap. 22, McGraw-Hill Book Company, New York, 1955.

E. C. Jordan, "Electromagnetic Waves and Radiating Systems," chap. 17, Prentice-Hall, Inc., Englewood Cliffs, N.J., 1950.

8-6 Two space stations in circular orbits at an altitude of 300 km are separated by 100 km. Booster rockets on both stations can be fired to keep the altitude and separation constant. Each station is equipped with two communication systems, one which operates in the frequency range 3 to 6 Mc and one which operates at 3 Gc. The antenna of the lower-frequency system is an infinitesimal dipole with a corner reflector, as shown in the figure. The microwave antenna is a paraboloid of revolution fed by a half-wave dipole which is a half wavelength from the origin of a corner reflector, as shown in the figure. The lower-frequency receiver in the space stations has a noise figure of 10 db and a 5 percent bandwidth. The microwave receiver has a noise figure of 3 db and a 0.1 percent bandwidth.

a Determine the magnitude of the far-zone electric field in the plane $\theta = 90°$ for the infinitesimal dipole in the corner reflector as a function of

$$kd = \frac{2\pi}{\lambda} d$$

Sketch the radiation pattern for maximum radiation in the $\phi = 0$ direction, when there is one lobe in the pattern and when there are three lobes.

b Determine the gain of the microwave antenna (in the direction of maximum radiation).

c One form of microwave communication which might be used requires that radio-frequency pulses be transmitted between stations. If the signal-to-noise ratio of the received pulse is to be 10 or greater, determine the transmitter power required at the other station. Approximately what length pulse should be used, if the received pulse is to have its maximum value R for at least one-half the pulse length?

d What transmitter power would be needed at a ground station to communicate with the space stations with a signal-to-noise ratio of 10? Assume the ground station has the same antennas as the space stations, which are passing directly overhead.

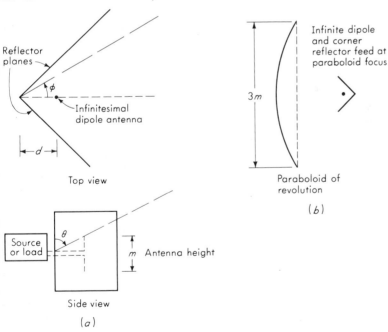

specific microwave communication systems

Could the ground station easily communicate with the space station using the lower-frequency system? What problems might occur?

e Design a simple system to monitor the distance above ground and the separation of the two stations. Briefly point out the problems that might occur and how your system avoids them.

f Suppose the critical frequency of the F layer were 5 Mc; discuss qualitatively the propagation between the two stations in the lower-frequency band (1) when the actual height of the F-layer maximum electron density is at 200 km and (2) when the actual height is at 300 km. (Use simple physical pictures to aid your presentation.)

8-7 In order to have uninterrupted communication with a spaceship that is going to explore the solar system, it is planned to have three earth-based stations which will transmit and receive 3-Gc microwave signals to and from a relay station on the moon. The relay station on the moon will convert the information onto a laser beam for transmission to the spaceship and will convert the laser-beam information from the spaceship onto the microwave carrier for transmission to the earth. In addition, so that all earth stations will know what the others are receiving, a telemetry synchronous satellite system will be set up to relay messages around the earth.

a Describe how the various stations should be oriented (i.e., sketch locations and the geometry).

b Draw a functional block diagram which describes the operations of the communication system. (First, draw a simple one. Then, as you work out the rest of the exercise, fill in more of the details.)

c For your earth-based microwave transmitters, you have basically only the following two choices for your output amplifier:
1. A multicavity stagger-tuned klystron amplifier with gain = 40 db, peak power output = 5 Mw, duty cycle = 0.001, bandwidth = 144 Mc.
2. A traveling-wave-tube pulsed amplifier with gain = 33 db, peak power output = 2 Mw, duty cycle = 0.002, bandwidth = 300 Mc.

Determine the channel capacities of these two transmitters relative to each other.

d Information will be placed on the microwave signal by frequency-modulating the backward-wave oscillator (BWO) which feeds the output amplifier. Determine the slowness factor (circuit velocity/velocity of light) and the pitch length of the slow-wave periodic circuit of the BWO so that a change in anode voltage from 200 to 300 volts will cover the desired bandwidth around 3 Gc.

e Determine the size of parabolic dish needed at the earth-based station and the dimensions and spacing of the moon-based rectangular array (an array was used on the moon because of its lighter weight) in order that your selected transmitter will provide a 100:1 signal-to-noise ratio after reception by a microwave maser whose effective noise temperature is 10°K. (Losses in the feed network of the array will be neglected.) Also state what aperture distribution you assume for the parabola and what type of elements is used in the array. *Note:* If you feel the sizes of any of the antennas are impractical, state how many output amplifiers could be paralleled at the transmitter to allow the use of an antenna of acceptable size.

f It is desired to have the moon radar automatically track whichever earth station is transmitting and switch when the earth stations switch. Draw a functional block diagram for automatically phasing the array, and suggest what components might do the various functions.

g The laser used for optical transmission is a pulsed ruby laser with peak power output = 5 Mw, duty cycle = 10^{-6}, and laser transition $\lambda = 0.6943$ μm. Neglecting losses in the modulator, what percent of the center frequency of the laser will be needed (assuming that the spaceship receiver operates at a signal-to-noise ratio of 10) to transmit the same channel capacity as is received at the moon?

h The largest useful transmitting lens is 200 in. in diameter, and, because of optical imperfections, this lens has a directivity which is equivalent to a 10-in.-diameter uniformly illuminated aperture. The spaceship detector uses a traveling-wave tube with a photocathode. This detector has a noise figure of 10 db. How far away can the spaceship be and receive signals from the moon with a 10:1 signal-to-noise ratio?

i The satellites in the telemetry relay system use solar batteries for power. A parametric difference-frequency amplifier whose gain is 18 db is used. However, this amplifier is used as an up-converter. What is its gain as an up-converter? What advantages are there in using it as an up-converter? A tunnel-diode oscillator is used to supply the pump power. Why was a parametric oscillator not used to supply the pump power?

j Describe other methods of solving this communications exercise and compare them.

REFERENCES

1 GOLDMAN, S.: "Frequency Analysis, Modulation and Noise," p. 83, McGraw-Hill Book Company, New York, 1948.

2 SKOLNIK, M. I.: "Introduction to Radar Systems," chap. 13, McGraw-Hill Book Company, New York, 1962.

3 KRASSNER, G. N., and J. V. MICHAELS: "Introduction to Space Communication Systems," McGraw-Hill Book Company, New York, 1964.

4 BALAKRISHNAN, A. V. (ed.): "Space Communications," McGraw-Hill Book Company, New York, 1963.

5 SCHLAGER, K. J.: Systems Engineering: Key to Modern Development, *IRE Trans.*, vol. EM-3, pp. 64–66, July, 1956.

6 FLAGLE, C. D., W. H. HUGGINS, and R. H. ROY (eds.): "Operations Research and Systems Engineering," The Johns Hopkins Press, Baltimore, 1960.

7 SIEGEL, K. M., H. A. ALPERIN, R. R. BRONSKI, J. W. CRISPIN, A. L. MOFFETT, C. E. SCHENSTED, and I. V. SCHENSTED: Bistatic Radar Cross Sections of Surfaces of Revolution, *J. Appl. Phys.*, vol. 26, pp. 297–305, March, 1955.

8 CRISPIN, J. W., R. F. GOODRICH, and K. M. SIEGEL: A Theoretical Method for the Calculation of the Radar Cross Sections of Aircraft and Missiles, *Univ. Michigan Radiation Lab. Rept.* 2591-1-H on Contract AF 19(604)-1949, July, 1959.

9 RIDENOUR, L. N.: "Radar Beacons," M.I.T. Radiation Laboratory Series, vol. 1, chap. 8, McGraw-Hill Book Company, New York, 1947.

10 BARTOW, J. E., G. N. KRASSNER, and R. C. RIEHS: Design Considerations for Space Communication, *IRE Trans. Commun. Systems,* December, 1959.

11 HAYDON, G. W.: Technical Considerations in the Selection of Frequencies for Communications with, via, and between Space Vehicles, *Natl. Bur. Std.,* Rept. 7250, December 1, 1962.

12 JENSEN, J., G. E. TOWNSEND, JR., J. KORK, and J. D. KRAFT: "Design Guide to Orbital Flight," McGraw-Hill Book Company, New York, 1962.

13 KENNAUGH, E. M., S. P. MORGAN, and H. WEIL: Scattering by a Spherical Satellite, *Proc. IRE,* October, 1960.

index

index

A (magnetic vector potential), 106
A (surface-wave attenuation factor), 165, 166
A_{eff} (effective antenna area), 139, 153
$\mathbf{a}_x$ (unit vector in x direction), 11
$\mathbf{a}_y$ (unit vector in y direction), 11
$\mathbf{a}_z$ (unit vector in z direction), 19
Absorption of radio-waves in atmospheric gases, 182–184, 217, 218
Adler, R. B., 214
Alperin, H. A., 239
Antenna arrays, 124–132
  array factor, 126
  nonuniform, 129, 130
    binomial, 130
    end-excited, 130
    optimum, 130
  uniform, 127–129
    broadside, 128
    end-fire, 128
Antenna equivalent circuit, 147, 151
  maximum power transfer, 147
Antenna impedance, 144–150
Antenna lens, 135, 136
Antenna noise, 211–213
  effective antenna-noise temperature, 211–213
Antenna pattern synthesis, 143, 144
Antennas, 103–161

Antennas, types of, 104
Aperture-type antennas, 135–142
  circular, 140–142
    beamwidth, 142
  effective area, 139
  fields, 137
  rectangular, 141
    beamwidth, 141
Arrays (*see* Antenna arrays)
ATR switch, 216–217
Atwood, S. S., 169, 185

B (magnetic flux density), 10
Bachynski, M. P., 185
Backscatter radar cross section, 221–229
  corner reflector, 222
  cylinder, 222
  flat plate, 222
  matched antenna, 221
  small aircraft, 222
  sphere, 222, 229
    method of calculation, 222–229
      high-frequency case, 226–229
      low-frequency case, 223–226
Bakanowski, A. E., 101
Balakrishnan, A. V., 214, 239
Bandwidth allocation, 2
Bartow, J. E., 240

Beam-coupling coefficient, 40
Bean, B. R., 185
Beck, A. H. W., 40, 41, 63, 74
Begg, B., 101
Bennett, W. R., 198n., 214
β (beam-coupling coefficient), 40
β (propagation constant), 10
β_e, 66
Biconical antenna, 148
Boltzmann's constant, *inside back cover*
Booker, H. G., 185
Bremmer, H., 185
Broadside antenna array, 128, 129
Bronski, R. R., 239
Buddenhagen, D. A., 102
Bullington, K., 164, 185
Bunching parameter, Klystron, 43
Burrows, C. R., 169, 185
BWA (backward wave amplifier), 22
BWO (backward wave oscillator), 22

C (traveling-wave tube gain parameter), 69
c (speed of light in vacuum), 11, *inside back cover*
Cahoon, B. A., 185
Cassegrain antenna, 154, 155
CCIR (International Telecommunications Union), 7
Circular loop antenna, 121, 123
 radiation pattern, 123
Circulator, 235
Coaxial line, 14, 15
Communication system, basic purpose, 1
 design, 5–7
 general, 2–4
 specific systems, 4
Conductivity of copper, *inside back cover*
Corner-reflector antenna, 134
Cosmic noise, 218
Crawford, A. B., 161
Crispin, J. W., 239
Critical frequency of ionosphere, 179
Cut-off frequency, 15
 for TE_{10} wave, 17

D (electric flux density), 10
Decoder, 3

Δ (phase angle of surface wave), 164–166
δ, 70
δ (loss angle), 167
Delta function, 203
Derosa, L. A., 214
Dipole antenna, 104–113
 field components, 110
 gain, 120
 potentials, 106, 109
 power outflow, 113
 Poynting vector, 112, 113
 radiation fields, 110
Direct wave, 164, 167–172
Directivity of antennas, 115, 116
Doppler frequency, 219
Down-converter, 87

E (electric field), 10
E′, 27
e (electron charge magnitude), *inside back cover*
e_0, 165
EHF, 7
Einstein, A., 98
Electric energy, 18
Electromagnetic energy propagation, 17–19
Electromagnetic fields, definitions, 10
Elemental dipole antenna (*see* Dipole antenna)
End-fire antenna array, 128
ϵ_0 (permittivity, dielectric constant), 10, 167, *inside back cover*
η_0 (intrinsic impedance), 11
η (electronic charge-to-mass ratio), 38

F (noise figure), 207–211
 excess noise figure, 208
Flagle, C. D., 239
Frequency-independent antennas, 156–159
 equiangular spiral, 158
 log-periodic, 159, 160
Frequency multiplexing, 2
Frequency spectrum allocation and designation, 7, 8
Friis, H. T., 149, 214

Fry, D. W., 161
Full-wave antenna (*see* Linear wire antennas)
FWA (forward-wave amplifier), 22

g (antenna gain), 116
Γ, 64, 65
Γ (reflection coefficient), 147
Γ (space-charge reduction factor), 204
γ (modulation factor), 88
Goldman, S., 201, 214, 239
Goodrich, R. F., 239
Gordon, W. E., 185
Goward, F. K., 161
Ground waves, 164–172
  direct wave, 164, 167–172
  reflected wave, 164, 167–172
  surface wave, 164–166

H (magnetic field), 10
h (Planck's constant), 94
Hahn, W. C., 35, 73
Half-wave antenna (*see* Linear wire antennas)
Haus, H. A., 214
Haydon, G. W., 240
Heffner, H., 101
Heil, A. A., 35, 73
Heil, O., 35, 73
HF (high frequency), 7
Horn antenna, 141, 142, 155
Huggins, W. H., 239
Hull, G. F., 161
Huygens-Fresnel principle, 135, 136

i (current density), 10
Image techniques, 132–134
  corner-reflector antenna, 134
Induced currents, multielectrode system, 27
  parallel plane diode, 25–27
  reflex klystron, 51
  two-cavity klystron, 44–47
Ionosphere, 176, 177
  effective dielectric constant, 176–178
  virtual height, 179, 180
  (*See also* Propagation of radio waves)

J (surface-current density), 13
$J_n(nX)$ (Bessel function), 46, 47
Jensen, J., 240
Johnson, H. R., 74
Johnson, J. B., 188
Jordan, E. C., 183, 236
JTAC (Joint Technical Advisory Committee), 7, 8

k (*see* Boltzmann's constant)
K band, 8
Keller, E. W., 214
Kennaugh, E. M., 240
King, R. W. P., 149
Klystron, 34–56
  reflex, 20, 48–55
    distance-time plot, 48, 49
    efficiency, 55
    mode number, 50
    power output, 55
    self-admittance, 51–53
  two-cavity, 35, 36, 41–48
    Applegate diagram, 36
    distance-time plot, 36
    transconductance, 44
Kork, J., 240
Kraft, J. D., 240
Krassner, G. N., 239, 240
Kraus, J. D., 161

L band, 8
Λ (array factor), 125, 126
λ_c, 17
Landon, V., 205, 214
Laser, 22, 98–101
  directivity of beam, 100
  R levels, 98
  ruby, 100
Lengyel, B. A., 102
LF (low frequency), 7
"Lighthouse" tube, 21
Linear magnetron amplifier, 22
Linear wire antennas, full-wave, 119
  gain, 121
  radiation pattern, 123
  half-wave, 119
  gain, 120
  radiation pattern, 123
Llewellyn, F. B., 73

McClung, F. J., 102
McLachlan, N. W., 74
Magnetic energy, 18
Maiman, T. H., 100
Manley, J. M., 85, 86, 92, 101
Manley-Rowe relations, 85, 86
Martindale, J. P. A., 161
Maser, 22, 93–98
  energy levels, 94–97
  uses, 98
Maxwell-Boltzmann equipartition, 195
Maxwell's equations, 10
Mei, K. K., 161
Metcalf, G. F., 35, 73
Meyer, J. W., 102
MF (medium frequency), 7
Michaels, J. V., 239
Microwave frequencies, amplification, 20
  designation, 8
  generation, 20
Microwave relay link, 4, 5
Millman, G. H., 185
Minimum detectable signal, 220
Moffett, A. L., 239
Morgan, S. P., 240
Mount, E., 101
μ (amplification factor), 31
μ (permeability), 10, *inside back cover*
MUF (maximum usable frequency), 179

n (unit vector in direction of normal), 12
Noise, 186–214
  antenna, 211–213
  characterization of noise sources, 189–199
  cosmic sources, 218
  current, 192
  electrical discharges, 186
  equivalent noise sources, 190
    current generator, 199
    voltage generator, 199
  excess, 208
  flicker, 189
  Johnson, 186
  manmade, 187
  N_0 (output power), 193
  shot, 189, 200
    basic noise event, 201, 202
    power spectrum, 202

Noise, sky, galactic, 187–188
  space-charge-limited diode, 204, 205
  thermal, 186, 194–200
  white, 186
Noise figure, 207–211
  cascade amplifiers, 209
Noise measure, 210
Noise model of communication system, 213–214
Norton, K. A., 164, 183

P (Poynting vector), 18
P band, 8
$p(f)$, 195
Paraboloidal reflector antenna, 135, 136, 154–156
  Cassegrain antenna, 154, 155
Parametric amplifiers, 82–93
  admittance matrix, 89
  down-converter, 87
  γ (modulation factor), 88
  physical principle of, 83, 84
  pump frequency, 86
  up-converter, 86, 89–93
    degradation factor, 92
    transducer power gain, 91, 92
  varactor diode, 90
Pederson, P. O., 185
Permittivity, 10
Peterson, L. C., 73
Phasors, 19
Φ (electric scalar potential), 106
Pierce, J. R., 36, 58, 64, 74, 214
Planck's constant, *inside back cover*
Plane waves, 10, 13, 14
  intrinsic impedance, 10
Poynting vector, 18, 112
  elemental dipole antenna, 113
  time-average value, 19
Propagation of radio waves, 162–186
  ground-wave set, 164–172
    direct wave, 164, 167–172
    reflected wave, 164, 167–172
    surface wave, 164–166
  sky wave, 176–182
    critical frequency, 179
    effective dielectric constant, 176–178
    ionized region, 176, 177
    maximum usable frequency, 179

Propagation of radio waves, sky wave,
    reflection from ionosphere, 178
    virtual height, 179, 180
  space wave, 167–172
    horizontal polarization, 168
    vertical polarization, 168
  tropospheric wave, 172–176
Propagation constant, 10
Pulse-repetition rate, 216
Pulsed radar system, 6, 7, 215–235
Pump frequency, 86

Q band, 8
Q_L (quality factor), 84

R (reflection coefficient), 165, 168, 169
R_r (radiation resistance), 149, 150
Radar, 215–222, 229–235
  cloud return, 233, 234
  cross section, 220
  duty cycle, 232
  equation, 219, 220
  frequency choice, 217
  maximum range, 220, 229–231
  maximum unambiguous range, 218
  pulse-recurrence-frequency choice, 234
  pulse-width choice, 233, 234
  radial velocity of target, 219
  range of target, 218
  scanning rate choice, 234, 235
  system diagram, 216
Radar cross section, 220–229
Radar equation, 219, 220
  maximum range, 220, 229–231
Radiation resistance, 149
Ramo, S., 19, 25, 73
Rayleigh-Jeans radiation law, 211, 212
Rayleigh scattering, 222, 223
Receiving antenna, 150–153
  maximum power transfer, 152
  open-circuit voltage, 151, 152
  of elemental dipole, 151
Reed, E. D., 101
Reed, H. R., 185
Reflex klystron (*see* Klystron)
ρ (volume-charge density), 10
ρ_s (surface-charge density), 13
Rice, S. O., 214

Ridenour, L. N., 240
Riehs, R. C., 240
Ronchi, L., 161
Rowe, H. E., 85, 86, 92, 101
Roy, R. H., 239
Rumsey, V. H., 161
Russell, C. M., 185

S band, 8
Saturation factor, 47, 53
Scattering of electromagnetic waves, 174, 175
Schelkunoff, S. A., 149
Schensted, C. E., 239
Schensted, I. V., 239
Schlager, K. J., 239
Schottky, W., 214
Schwartz, M., 214
Shannon, C., 2, 8
Sharpless, W. M., 102
SHF (superhigh frequency), 7
Shockley, W., 25, 73
Siegel, K. M., 239
Siegman, A. E., 95n.
σ (conductivity), 18
σ (radar cross section), 220–229
Silver, S., 160, 161
Singer, J. R., 102
Skolnik, M. I., 239
Sky-wave propagation, 176–182
Smith, G. F., 102
Smith, R. A., 161
Solid state devices, 12, 75–102
  (*See also under* specific devices)
Spangenburg, K. R., 36, 74
Spot noise figure, 207
Stegen, R. J., 161
Surface current, 13
Surface wave, 164–166

TEM wave, 12
TE_{mn} waves, 15, 16
TE_{10} wave, 16
Terman, F. E., 183, 236
Thayer, G. D., 185
θ_g (gap transit angle), 39–40
θ_o, 51
Tien, P. K., 101

Toraldo di Francia, G., 161
Townsend, G. E., 240
TR switch, 6, 216–217
Transit angle of gap, 39
Transit-time effects, 23–34
  diode, 31
  space-charge-controlled tubes, 28–34
Transmission channel, 3
Transverse electric waves, 15
Transverse magnetic waves, 15, 17
Traveling-wave tube (TWT), 22, 56–74
  bunching, 57
  C (gain parameter), 69
  circuits, 60–63, 66, 67
    Brillouin $\omega\beta$ diagram, 60, 61
    equivalent, 66, 67
    helix, 60
  efficiency, 59
  electronic tuning, 63
  gain parameter, 69
  linear magnetron, 58
  power gain, 72
  propagation constant, 61
  spatial harmonics, 62, 63
Triode, 28–34
  amplification factor, 31
  common-cathode, 30
    input impedance, 30
  common-grid, 28, 29
    input impedance, 30
  input admittance, 32–34
    variation with transit angle, 32, 34
  transadmittance, 31
Tubes, fundamental limitations, 22, 23
  high-frequency limitations, 23–25
Tunnel diode, 76–82
  amplifier, 80–82
    gain, 81
    gain-bandwidth product, 81
  characteristic iv curve, 76, 77
  conductance, 78
  equivalent circuit, 79, 80
Two-cavity klystron (*see* Klystron)
TWT (*see* Traveling-wave tube)

Uenohara, M., 101, 102
UHF (ultrahigh frequency), 7
Uhlir, A., 101
Up-converter (*see* Parametric amplifiers)

V band, 8
Van der Ziel, A., 189n.
Van Duzer, T., 19
Van Vleck, J. H., 185
Varactor diode, 90
Varian, R. H., 35, 73
Varian, S. F., 35, 73
VHF (very high frequency), 7
Vincent, B. T., 101
Virtual height of ionosphere, 179, 180
Visibility factor, 231
VLF (very low frequency), 7
Von Aulock, W. H., 161

Wade, G., 101
Wave equation, 10
Waveguides, 15–17
  coaxial, 14, 15
  rectangular, 15–17
    cut-off frequency, 15
Weil, H., 240
Western Electric 416, 20
Whinnery, J. R., 19
White noise, 205
Wire antennas, 116–123
  fields, 118
  radiation patterns, 123
Woodward, B., 161

X (bunching parameter), 43
X band, 8

Z_o (characteristic impedance), 68
Z_{TE}, 16